648 Billion Sunrises

Two quotes capture the essence of this book. The first tells of the excitement and impact that a life in geology can have; the second summarises a plea that this book attempts to answer.

Sunrise, wrong side of another day,
Sky high, and six thousand miles away
Don't know, how long I been awake
Wound up, in an amazin' state.

Taken from the song 'Motörhead' by the band Motörhead from the album Motörhead. *Lyrics by Ian Frazer Kilmister, better known as Lemmy. EMI United Partnership Limited.*

'Oh rocks – tell us in plain words.'

Molly to Leopold Bloom in James Joyce's Ulysses.

648 Billion Sunrises

A Geological Miscellany of Ireland

Patrick Roycroft

Orpen Press

Published by
Orpen Press
13 Adelaide Road
Dublin 2
Ireland

email: info@orpenpress.com

www.orpenpress.com

1 3 5 7 9 10 8 6 4 2

Paperback ISBN 978-1-909895-68-3
ePub ISBN 978-1-909895-69-0
Kindle ISBN 978-1-909895-70-6
PDF ISBN 978-1-909895-71-3

A catalogue record for this book is available from the British Library.

Printed in Dublin by SPRINTprint Ltd.

Illustrations

The following plates are between pages 90 and 91

Plate Figure 1. The Irish flag – Nature's version. Left: green, white and orange as displayed by the three mica minerals of chlorite, muscovite and biotite. This 'flag', only 1.5 mm across, was found in a thin section of Leinster Granite from south Dublin and photographed in polarised light using a geological microscope. (The ancient Celtic text *The Annals of the Three Micas* says that the green, white and orange are the sept colours of Phil O'Silicate, a sheet-maker to Rí Roc na h'Éireann (King Rock of Ireland). The normal version of the Irish flag is included (right) for comparison.) See Chapter 2. Photo: Patrick Roycroft.

Plate Figure 2. The Blue Waterfall. Cavers were stunned by this solid 'waterfall' formed from a variety of copper minerals, including azurite, when they came across it while exploring the now-abandoned underground copper mines at Tankardstown, Bunmahon, County Waterford. Man and Nature have unwittingly worked together to produce something really beautiful here. See Chapter 3. Photo: Martin Critchley.

Plate Figure 3. World-class folds at Loughshinny, north County Dublin. The many grey layers you see are limestones and mudstones that were originally deposited horizontally and underwater during the Lower Carboniferous. 'Ireland' at the time was at the equator. But geologically soon after their formation, some 300 million years ago, these limestones and mudstones suffered the indignity of Iberia and North Africa obliquely ramming into Ireland from the south, causing the rocks that are now at Loughshinny to spectacularly buckle in a distinctive zigzag way. See Chapter 5. Photo: Patrick Roycroft.

Plate Figure 4. To quote a song from *Sesame Street*: 'One of these things is not like the others.' Above: The brooding Great Sugar Loaf of north Wicklow. It is made of the metamorphic rock quartzite, itself derived from originally pure sands that, about 550 million years ago, slumped en masse into a sedimentary basin; these sands were then metamorphosed to quartzite 400 million years ago, then exposed to erosion within the last 20 million years to produce the shape we see today. Below: Conchagua stratovolcano, El Salvador. This is a real volcano, but geologically very different to the Great Sugar Loaf. See Chapter 6. Photos by Patrick Roycroft (Sugar Loaf) and Rick Wundeman (© Smithsonian Institution).

Plate Figure 5. Top left: part of the real thigh bone of the herbivorous *Scelidosaurus*. Top right: artist's impression of *Scelidosaurus*. Bottom left: part of the real thigh bone of the carnivorous *Megalosaurus*. Bottom right: artist's impression of *Megalosaurus*. Ireland's only two dinosaur bones. Both bones were found in marine sedimentary rocks from the Jurassic Period, loose on the beach at Islandmagee, County Antrim. See Chapter 7. Photos of bones by Mike Simms; *Scelidosaurus* from Wikicommons © Arthur Weasley; *Megalosaurus* from Wikicommons © Mariana Ruiz Villarreal (as altered by FunkMonk and Steveoc86).

Plate Figure 6. The author and Irish gold. Left: I have just found my very first fleck of gold – the tiny speck in the middle of the black rectangle of insulating tape – from the River Dodder at Bohernabreena, south Dublin. Right: the author holding a replica of the famed Great Wicklow Nugget (TCD collection), the largest gold nugget ever found in Europe. The original nugget was given to 'Mad' King George III of England, who seems to have had it melted down and made into a snuffbox. See Chapter 9. Photo (left): unknown; photo (right): Patrick Wyse Jackson.

Plate Figure 7. Cotterite is an extremely rare form of the otherwise extremely common mineral quartz. It is characterised by having a curious pearly metallic lustre on its outer surfaces, and there are only about 20 specimens in the world. Cotterite is named after a 'Miss Cotter' from Rockforest, near Mallow, County Cork, and almost all known specimens come from this locality. The specimen here is in University College Cork and is the biggest cotterite of them all. See Chapters 10 and 16. Photo: Richard Unitt.

Plate Figure 8. Huge amethyst group from Cork. This magnificent specimen, held by Trinity College Dublin, was collected by Robert Ball, the first director of the Dublin University Museum, from a rich seam of very large and deep-purple amethysts that once occurred in the Blackrock area of Cork city. See Chapter 11. Photo: Patrick Roycroft.

Plate Figure 9. Zoned muscovite and sparkling granite. Above: a large, 2.5cm-across, beautifully zoned crystal of muscovite taken from sparkling muscovite-rich granite near Three Rock Mountain (Walshe's quarry in Stepaside, south Dublin) and photographed under the crossed polarisers of a petrological microscope. Below: the glittering variety of Leinster Granite from which this muscovite comes. Note: all the muscovites are seen to be fabulously zoned when viewed in crossed polarised light. See Chapters 12 and 13 and this book's front cover. Photos: Patrick Roycroft.

Plate Figure 10. *Urocordylus wandesfordii*. Left: A drawing of the fossil of one of the largest early amphibians, about 310 million years old, to come out of the Upper Carboniferous Jarrow Coal Seam within the Castlecomer Plateau, County Kilkenny. This is an important representative of the early amphibians, the first vertebrates to conquer land. Right: A reconstruction of this creature. See Chapter 13. Photo: Patrick Roycroft; artist's impression from Wikicommons © Smokeybjb.

Plate Figure 11. Giant Irish deer. Left: a complete giant deer skeleton on public display in the foyer of the Museum Building, Trinity College Dublin. It was originally found in a bog at Lough Gur, County Limerick. Right: in the hallway of the Royal Society of Antiquaries of Ireland (Dublin) is a giant deer skull mounted at natural height; and that is me, feeling dwarfed, looking up at it. See Chapter 13. Photos: Patrick Roycroft (TCD deer); Niamh McCabe (RSAI deer).

Plate Figure 12. *Cooksonia*. The very earliest plants that grew on land: found around the Devil's Bit Mountain, County Tipperary, and dating from 425 million years ago (Silurian Period). Small stems can be seen to end in bulbous sporangia (organs for reproduction). In a sense, all land plants, from delicately scented roses to Giant Redwoods, trace their ancestry back to these Tipperary fossils. See Chapter 13. Photos: John Feehan; cartoon reconstruction © Smith609, Wikimedia Commons.

Plate Figure 13. *Siphonophyllia samsonensis*. This is an enormous solitary coral from the Lower Carboniferous and was collected as a loose specimen on the beach at Serpent Rock, County Sligo; it was not hammered out: this is a protected site. One can see hundreds of these huge animals at Serpent Rock entombed in the muddy limestone bedrock that was once a tropical sea floor and is now a temperate seashore. See Chapter 13. Photo: Patrick Roycroft.

ACKNOWLEDGEMENTS

It is a pleasure to thank the following people. First and foremost, both my parents, Patrick (RIP) and Jennifer, who not only indulged but actively encouraged my enthusiasm for geology and who put up with countless boxes and bags of rocks about the family home for years: every time we moved house, new collecting opportunities opened up. Catherine McAuley for being that special person in my life. My very good friends Derek Sargent, David Edwards, Tom Dwyer, Ray Flynn and Marianne Osborn. Many thanks go to my former PhD supervisor, Pádhraig Kennan, whose advice and knowledge was then, and is now, priceless. Mary Mulvihill (RIP) for enjoyable employment with Ingenious Guides Ltd and sage advice about writing a book. Thanks to Trevor Sargent and to former Orpen Press commissioning editor Elizabeth Brennan for initially kick-starting this project, and to current Orpen Press commissioning editor Ailbhe O'Reilly, and to editors Eileen O'Brien and Fiona Dunne for their care and patience in seeing it through. Thanks also to cover designer Clifford Hayes, and to Jim Shine for the index.

Thanks to all those who kindly gave me permission to reproduce figures or quotes: Dunedin Press, Elsevier, the Geological Survey of Ireland (for maps 1.2 to 1.7), and Motörhead (via EMI). Thanks to everyone who let me use their photographs (credits given with pictures). For additional references, technical comment and photographs, thanks to Chris Standish (University of Southampton), Patrick Wyse Jackson (Trinity College Dublin), Matthew Parkes (National Museum of Ireland), Michael Simms (Ulster Museum), and Peter Haughton (University College Dublin) for the Haughton genealogy. To John Kennedy (UCD) for computer help. To Stephen Moreton (chemist with PQ Corporation, UK) for invaluable, first-hand information on Irish minerals. John D'Arcy of Lapis Jewellers (Dublin). Mark Hennessy, TCD Geography Department. Thanks also to genealogist Brian Donovan (Eneclann) and all the Findmypast.ie team for facilities and advice.

And thanks to my brothers, Niall and Brendan, and my sister, Sian, and their families; Aunt Joyce; and all my various cousins and other relatives who have taken an interest. Thanks to all my friends and mentors over the years at Trinity College Dublin; at University College Dublin; at the CRMC2 laboratory in Marseille (especially Professor Alain Baronnet); and colleagues who worked with me at the now defunct H.W. Wilson Company (Dublin office). And thank you to all the staff at the National Museum of Ireland – Natural History for help with the UCD mineral collection project. Apologies to anyone left out: you'll get into the second edition! The publisher has made every effort to contact all copyright holders but if any have been overlooked, we will be pleased to make any necessary arrangements.

I dedicate this book to Mother Nature for putting together the extraordinary land that is Ireland.

Biographical Note

Patrick Roycroft was born in Woodstown, Waterford (Ireland), in 1965. But it was in Galway in 1972 that he was bitten by the geology bug *Geomania infestans*, and has remained happily infected ever since. Patrick did his undergraduate degree in Trinity College Dublin, graduating in 1987 before heading to University College Dublin to complete a PhD in geology under the 'super vision' of Dr Pádhraig Kennan. A physically monstrous, but very colourful, thesis on the world-class muscovites of the Leinster Granite was completed in 1995. This was followed by a European Union Marie Curie Fellowship to Marseille (France) with Professor Alain Baronnet where, between 1995 and 1997, he extended his work on the growth and corrosion patterns in natural micas. He was subsequently awarded a 'rarer than hen's teeth' second European Union post-doc to study again in UCD. From 2000, however, full-time geology had to be put aside, and he worked as an editor for the H.W. Wilson Company, an American publishing house that went bust in 2011. He then enjoyed working with Mary Mulvihill's Ingenious Ireland, both as a tour guide and a researcher; sadly Mary passed away shortly before this book was completed. He has also been awarded two Heritage Council grants (2014 and 2015) to work with the National Museum of Ireland – Natural History in saving UCD's massive, but unknown, mineral collection. And in January 2015, he was appointed copy editor with the geological journal *Elements*.

Contents

Preface

This book is a slightly mischievous miscellany of Irish geology written for the layperson and anyone with an interest in Irish natural history. No prior knowledge of geology is assumed.

I have taken the concept of a miscellany – bringing together a variety of interesting things and themes – and then raised the stakes: I have also included some original research, so the reader, whether new to Irish geology or a grizzled veteran, will find some material herein that cannot, at the time of writing, be found elsewhere. There are also elements of personal biography for that often-lacking human touch: geology is a human endeavour, after all. Dotted about are a few suggestions for small projects or things one could do oneself. And there is the occasional stylistic literary twist in which a subject is dealt with in a non-conventional way, e.g. tabloid newspaper, radio monologue and Biblically.

Geology has the slightly undeserved reputation of being 'a bit difficult'. But it really isn't. Perhaps people's trepidation is because it can be problematic to emotionally bond or behaviourally relate to rocks: under normal circumstances, rocks don't move, they are not alive, they don't do cute and cuddly things, you can't eat them, and they can't immediately communicate with you. At first sight, rocks don't have the qualities that most people need to strike up a relationship.

But rocks *are* trying to communicate. They expose themselves to us all the time and beg us to look at their exposures!

We just have to know how to read their communication signals. Fossils, lest we forget, were once real living, respiring, reproducing, defecating creatures, some of which you could have eaten, and some of which could have eaten you. Sedimentary rocks are trying to tell you of real past geographical environments, and when you can listen to their story it is like seeing in your head one of those glossy coffee-table travel books full of photos of tropical islands, stormy seas, placid lakes, sandy deserts, sweaty jungles or Polar blizzards. Metamorphic rocks are whimpering or screaming of the thrills and horrors of strange subearthly transformations: a geological form of going from (nice) Dr Jekyll to (gneiss) Mr Hyde. And igneous rocks tell of goings-on in the hidden world of the pulsing red glow of Earth's deep infernal interior.

Rocks should be thought of as our 'Tribal Elders' who know the stories of how we came to be. They remember their stories well over many, many years; and their stories are thrilling. But with increasingly immense age even they can sometimes forget, and their speech becomes slurred. Geologists are the initiates who can best understand our Venerable Rock Elders: geologists do their best to translate the Rock Elders' stories from the old tongue ('Lithican' – with its myriad dialects and not fully understood grammar) into tales that everyone can understand.

Each of the essays in this book can be read independently of the others; they can be read in any order; and where the pieces themselves are multipart, they can be dipped into at random. Nevertheless, I suggest reading the first essay (Chapter 1) first because this is the one that offers the broad overview of Ireland's entire extraordinary history and, hence, if you are new to geology, lays the foundation for so much else. Should you wish to read the book linearly from start to finish, there is a tripartite structure designed to facilitate this: a series of short 'rocky' essays to start off with; a middle section of longer essays that often comprise many bite-sized bits for easy digestion; and a final section dealing with the more human aspects of geology. And the whole set opens and ends with the same theme – the whole geological saga of Ireland – written in very different ways.

The reader will, at some point, notice that I like music. This book is musical. It is constructed as if in a rough 'sonata form'. There is the exposition of the main theme at the beginning (the geological history of Ireland, played 'straight'); the development section in which aspects and parts of that theme are chosen for development in a myriad of ways; and there is the final recapitulation of the theme when, after having experienced all its adventures, we read the theme anew, in a fresh and transformed way. And within the body of the development section, individual essays sometimes have their own 'musical' substructures, such as the Scherzo and Coda of Chapter 6, or the Introduction, Prelude and Fugue that is so appropriate to Chapter 10.

Above all else, this book is about enjoying many of the unique aspects of Irish geology. For a relatively small island, we have an enormous diversity of rocks, both in type and in age. Irish rocks, minerals and fossils have played an often pivotal role in the history of geological science. And research done in Ireland on Irish rocks still has the potential to produce scientific paradigm shifts. In short, we have some amazing geology that we can all be proud of. This book sets out why and does so in a way that is not always conventional, but is, I hope, always enjoyable.

1

648 *Billion Sunrises: From the Beginning*

Ireland's Very Ancient History

Preamble

What better way to start a geological miscellany of Ireland than to give a concise, up-to-date account of Ireland's entire history as the first chapter? This addresses two challenges simultaneously: to give a not-too-lengthy account of how Ireland came to be Ireland, and to provide a broad framework that the reader can use when reading other chapters. We learn in school about the Céide Fields in Mayo; the High Kings of Ireland; the invasions by the Vikings, the Normans and the Elizabethans; the rebellions and the uprisings; and the story of Ireland becoming an independent nation. Our longer geological history is but an extension backwards of all that. It is still our history. Every bit as fascinating, eventful, and, at times, almost unbelievable.

Introduction

A few points will be helpful to know before we start. The Earth's crust now, and for most of its 4,567-million-year history, is made of two types. First, continental crust, which is relatively light and dominated by rocks made of silicon and aluminium minerals. Second, oceanic crust, which is relatively heavy and dominated by iron and magnesium minerals. The crust is cracked, like a roughly cracked egg, and thereby divided into plates. Very crudely, one could say that continental crust and oceanic crust form their own plates. Unlike an egg's shell, these plates can move around the surface of the Earth, driven by slow, powerful convection currents beneath them in Earth's mantle (imagine an egg's albumin convecting and moving the cracked shell fragments above it). This movement causes the plates to do one of three things: split further apart, rub past one another, or crash into each other. Every 600 million years or so, either all or most of the continental parts of our shell seem to come together to produce a massive supercontinent, which, after a while, itself breaks apart into new continents. These new, more normal-sized continents go for a wander about the globe until, a few hundred million years later, they all bang into each other again and form another supercontinent. This idea of periodic supercontinent formation and break-up is much debated in geology. At the moment, geologists think that this vast cyclical process has happened four, possibly five, times during Earth's entire history.

The edges of continents are where things usually happen, and there are two broad types of continental-edge processes that are important for our story. The first is that the edges of continents are where lots of sediments tend to accumulate. Thus, we get limestones and coral reefs, large river-delta deposits, and sands and muds from the continental shelf falling down the continental slope and accumulating in layers on the deeper ocean floor. So, the seaward margins of continents are where there is a great deal of sedimentation going on; this can continue for many millions of years, building up thick piles of limestone, sand and mud in the relatively shallow waters along the coast of a continent.

The second process is that when an ocean closes – or, to put it another way, when two or more continents are on their way to having a slow-motion collision with each other because there tends to be a continent at the edge of every ocean – the heavier (iron-rich) oceanic plate is pushed under the lighter (silica-rich) continental plate and down into the Earth's mantle. This process is called subduction, and it is a bumpy ride. Because subduction happens to hard, irregular masses of rock, there is friction on the underside of the continental plate above the down-going oceanic plate and this, for a short while, causes the margins of that continental plate to extend, stretch and split. And where crust is pulled apart, it is thinned and drops down a bit, resulting in widespread subsidence. As a consequence, we get large depressions called sedimentary basins on continental margins where subduction is taking place, and these basins fill up with sediment. If the stretching continues to the point where cracks reach down to the mantle, then volcanic rocks are also introduced.

This is important to remember because, when the oceanic plate is just about gone, the two continents on either side meet and collide and all the different types of sediments that had accumulated on both the respective continents' margins will now get squashed and metamorphosed. Mother Nature literally plasters these sediments and volcanic rocks onto the hard outer parts of the older, pre-existing continents (cratons) of which they were originally at the edge. The zone of collision where continents finally meet is called, not unreasonably, a suture zone. Large amounts of metamorphosed and structurally mangled rock are formed. Mountains are formed.

After a long while, such an amalgamated continent (formed from the two or more continents that have collided and welded themselves together) will split apart in a new way. A new ocean will form between them and get wider, and these newer continents will, in turn, wander off and accumulate marginal sediments of their own until, one day, the ocean that spread them will contract and these 'new' margin sediments (and volcanics) will, ultimately, get metamorphosed and plastered onto *their* old continental hosts (which now, of course, will include any previous round of squashed rocks). And so it goes on.

Each time a new lot of sediment gets metamorphically and structurally plastered onto an old continent from a continental collision, the whole process also affects, to some extent, the previous generation(s) of plastered, metamorphosed sediments and volcanics. And this is how many really old rocks get so very messed up and complicated-looking. Mash up some sediments, volcanics and igneous rocks in two or three different continental crashes over a billion years or more and it becomes a virtuoso task for any geologist to unravel.

Some knowledge of this grand Earth supercontinent cycle is helpful when trying to visualise the large-scale events that happened in the formation of 'Ireland' (only *very* recently the island that we know and love today, hence the inverted commas).

It is the key that allows us to see the bigger picture and to make sense of our complicated geological map.

Do you, by any chance, recall the film *The Curious Case of Benjamin Button*, the 2008 fantasy drama directed by David Fincher and starring Brad Pitt and Cate Blanchett? In it, the main character, Benjamin Button, who is born very elderly and on the verge of death, gets younger and younger until, at the end of his life, he is an infant. Geologists love an analogy, and here's one to help you comprehend how to 'visualise' Ireland's rocks. *As we see them now*, rocks are analogous to the Benjamin Button character. The rocks that were around when Ireland was at its very youngest (back in the Paleoproterozoic Era) are, as we see them today, the oldest and gnarliest, whereas rocks formed during the Pleistocene Epoch (after Ireland had been around for well over 1,700 million years) look the youngest. Thus, when we consider how the rocks looked when Ireland was born and how the rocks look today, we really do have a geological case of Benjamin Button. The mind-bogglingly old, wrinkled, deformed first rocks are actually Ireland's baby pictures, with its face screwed up, a dirty nappy, and bawling lustily. And the newly formed rocks of the last few thousand years are how the newest Ireland looks after the best part of one and three-quarter billion years ... or about 648 billion sunrises.

Shakespeare gave us the seven ages of man. Today, we compartmentalise our lives into ages that include 'infancy', 'childhood', 'adolescence', 'middle age' and 'old age'. Similarly, we can subdivide 'adolescence' into all the individual teen years, and subdivide years into months, months into weeks, weeks into days, days into hours ... and so on. Exactly the same is done for Earth's geological ages, and we can keep track of Ireland's history using these. Broadly speaking, Ireland does not have seven ages but fifteen. These, and their actual times in millions of years ago (abbreviated as 'Ma'), are as follows:

1. Paleoproterozoic (2,500–1,600 Ma)
2. Mesoproterozoic (1,600–1,000 Ma)
3. Neoproterozoic (1,000–542 Ma)
4. Cambrian (542–488 Ma)
5. Ordovician (488–444 Ma)
6. Silurian (444–416 Ma)
7. Devonian (416–359 Ma)
8. Carboniferous (359–299 Ma)
9. Permian (299–251 Ma)
10. Triassic (251–200 Ma)
11. Jurassic (200–145 Ma)
12. Cretaceous (145–65 Ma)
13. Paleogene (65–23 Ma)
14. Neogene (23–2.65 Ma)
15. Quaternary (2.65 Ma to now)

Just to clarify – nobody refers, for example, to the Neogene as Ireland's fourteenth age. The numbers are only for convenience here.

Before reading on, it would be useful to have at the ready the geological map of Ireland and the stratigraphic column that are reproduced at the back of this book. Now brace yourselves! The following synopsis of Ireland's entire history is going to be a roller-coaster ride through time and space.

Our 648 Billion Sunrises

We can somewhat romantically say that Ireland was 'born' 1,780 million years ago on the little island of Inishtrahull, off Malin Head, County Donegal. Or, more correctly, the northwest (NW) half of Ireland was born here. The southeast (SE) half was not yet born. (*Reader's voice*: 'Hang on, *what*?!')

Okay, there's one other basic fact that you need to know about the land that makes up Ireland. It has actually spent most of its life as two quite separate halves. Long before Ireland was an island, long before we were even a coherent landmass, Ireland, for most of its existence, was a divided land – literally! Imagine a diagonal line running from Drogheda (County Louth) in the NE through to the Shannon Estuary (counties Clare/Limerick) in the SW separating Ireland into two oblique halves. Now insert an entire ocean between these two halves, and that is how Ireland was until about 400 million years ago (400 Ma) when there was a humongous geological Act of Union. Before 400 Ma, the two parts – NW Ireland and SE Ireland – led completely separate lives. So, in a sense, Ireland was born as twins who then became conjoined later in life. And for the baby pictures, we have to treat each twin separately.

So, let's take it from the top.

Northwest Ireland was 'born' in the Paleoproterozoic Era, some 1,780 million years ago, on Inishtrahull. The rocks on this little island are the oldest rocks we have, and the 'midwife' who determined their age was Stephen Daly of University College Dublin. He and two colleagues made this remarkable discovery in 1991. The rocks themselves are igneous rocks called syenites (think of granite but without the glassy quartz), and they were really put through the geo-wringer between the time they originally crystallised and now. Today, they have been thoroughly metamorphosed into a rock type called a gneiss (pronounced 'nice', a type of metamorphic rock that has been subjected to lots of heat and shearing). These gneisses would, presumably, have originally been part of Earth's (postulated) second supercontinent of Nuna – the name is a quasi-acronym based on 'Northern Europe North America' – which was assembling at around 1,700 Ma. (Aside: the first postulated supercontinent formed maybe 2,700 Ma before any part of Ireland was even a twinkle in Mother Nature's eye.) Some have speculated that during this very early time, NW Ireland was in the northern hemisphere, at about 40 degrees north (see Figure 1.1), and starting to head south as part of a landmass called Laurentia – of which more anon. Parts of Australia at this time were possibly geographically *above* us, maybe close to the North Pole. The past is a different country, alright.

The next-oldest rocks we have also come from NW Ireland: these are the Annagh gneisses, which comprise much of the rocks at Annagh Head and around Blacksod Bay and the Mullet Peninsula in County Mayo. These are also Paleoproterozoic to Mesoproterozoic in age, ranging from 1,750 Ma to 1,180 Ma, and were also once mostly igneous rocks that have been severely heated and squashed several times.

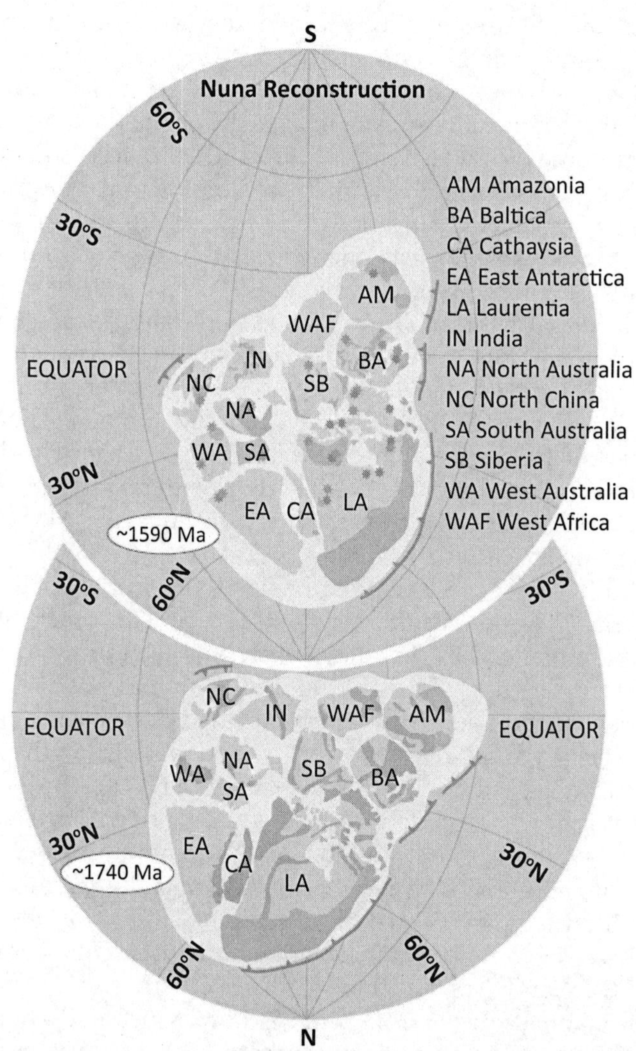

Figure 1.1. Map of the world, which then comprised only the Nuna supercontinent, at one and three-quarter billion years ago and one and a half billion years ago. At this time, only the oldest parts of northwest Ireland existed and they were probably joined to Laurentia ('LA' on the maps). This is the birth of Ireland – and it happened almost where we now are today (note the latitudes): it has taken us almost two billion years to get back to where we were 'born'. From Zhang et al. (2012), with permission from Elsevier.

The birth of SE Ireland took place in Rosslare, County Wexford, possibly during the Neoproterozoic Era or even the Mesoproterozoic. Nevertheless, the rocks themselves were very probably born in the southern hemisphere ('geographically opposite' and a little after the oldest rocks of NW Ireland) and are also gneisses, but these gneisses were originally sediments and volcanics, as opposed to igneous intrusions. These first SE Ireland rocks were part of a continent that was once attached to what is now Africa and South America. We know that these rocks were involved in a major early continental collision, called the Grenville orogeny. The word 'orogeny' just means 'mountain-building event' and each one that geologists have identified has been given an appropriate name, in this case 'Grenville' after Grenville, Canada. The Grenville orogeny happened some 1,000 million years ago. This means that Rosslare was part of Earth's third supercontinent, called Rodinia, which started to assemble itself around 1,300 Ma and was still going on at around 1,000 Ma.

By about 750 Ma, the Neoproterozoic supercontinent of Rodinia, having had a good existence, was itself about to break up into smaller continents. NW Ireland, which had been in the northern hemisphere, was now almost at the South Pole and was attached to a continent called Laurentia, which included parts of what is now North America and Greenland. SW Ireland was also in the southern hemisphere, but not in close proximity to NW Ireland.

Rodinia, like the previous Nuna supercontinent, broke apart over the next few hundred million years into new, mostly normal-sized continents, and the two parts of Ireland, still separated, were attached, once again, to two quite separate chunks of plate. NW Ireland was now at the southern edge of the Laurentia continent, while SE Ireland was attached to the northern edge of a much smaller snapped-off bit from what is now Africa and Amazonia and was part of the very large southern continent of Gondwana (termed the fourth supercontinent by many). This snapped-off bit, containing SE Ireland, was the micro-continent of Avalonia. Separating the two parts of Ireland – attached to Laurentia and Avalonia, respectively, both of which were still in the southern hemisphere – were the waters of the Iapetus Ocean (see Figure 1.2). We can imagine the Iapetus Ocean lapping the shores of both NW Ireland

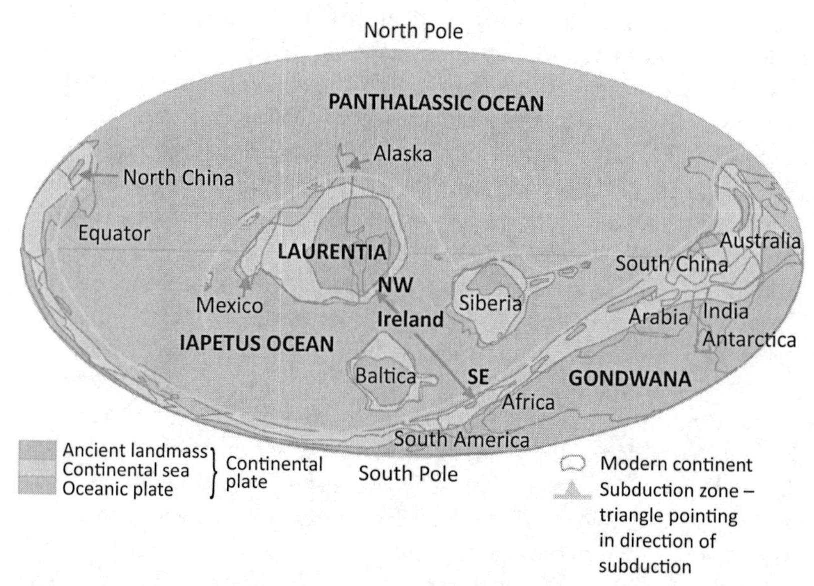

Figure 1.2. Map of the world during the Cambrian Period (542–488 million years ago).

and SE Ireland, albeit the two parts were still some hundreds (maybe a thousand) kilometres apart.

Margins are where most geologically interesting events occur. During the Neoproterozoic Era and into the Cambrian Period, a thick pile of sediments and volcanics accumulated geographically seaward of (and off) the oldest parts of NW Ireland, and off the whole of the Laurentia continent more generally. NW Ireland (Laurentia) had turned partly around and was now heading north towards the equator (although not there yet). Limestones were being laid down because we were in warmish waters – NW Ireland was maybe 30 degrees south of the equator during the Early Cambrian. However, SE Ireland was not: it was still hundreds of kilometres further south, possibly at 70 degrees south, and much colder. Nevertheless, it is during the Cambrian Period that rocks such as those that now form Howth (County Dublin) and Bray (County Wicklow) were deposited off the shores of old SE Ireland (as part of Avalonia).

During the Ordovician Period and the Silurian Period, the Iapetus Ocean, which had opened as a result of Rodinia splitting, was now inexorably closing (see Figure 1.3). As it did so, deep-water sediments, including the metal-rich products of underwater black smokers, formed off the northern end of SE Ireland. And, as the ocean itself contracted, more volcanic rocks erupted, forming the swath of Ordovician volcanics now seen in counties Waterford, Wexford and Wicklow, as well as the Lambay Island volcano and a big volcano now partly exposed at Balbriggan in north County Dublin.

What happened next was the most momentous event in Ireland's history. Two became one. It was the Caledonian orogeny – the true Act of Union that physically joined NW Ireland to SE Ireland. The two halves of Ireland, one on the Laurentian margin and the other on the Avalonian margin, finally met in an almighty geo-crunch as the Iapetus Ocean disappeared and the two continents on either side collided during the early Devonian Period, some 400 million years ago (see Figure 1.4). The sediments that had accumulated on the seaward margins of NW Ireland, called the Dalradian sediments (which today form large parts of counties Galway, Mayo and Donegal), got folded, thrusted, buried, metamorphosed,

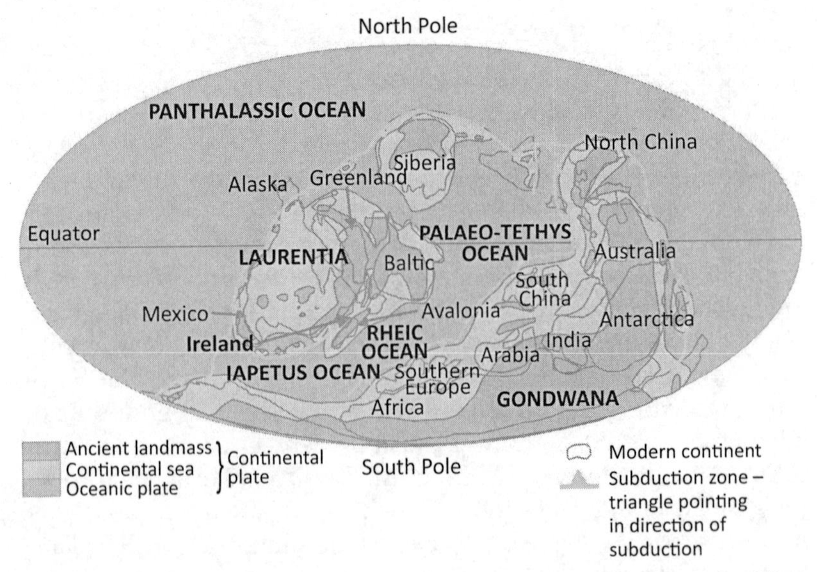

Figure 1.3. Map of the world during the Silurian Period (444–416 million years ago).

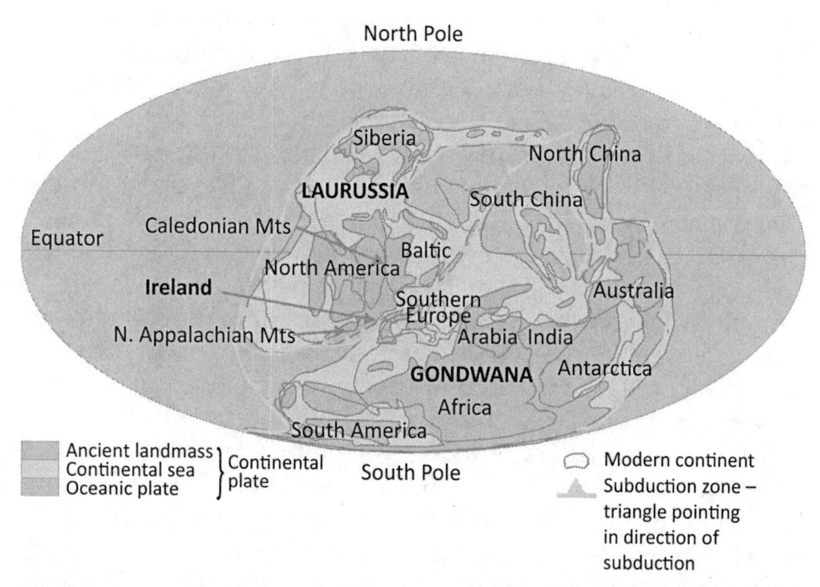

Figure 1.4. Map of the world during the Devonian Period (416–359 million years ago).

deformed and twisted. They effectively got plastered onto the older Irish rocks that had formed during the Paleoproterozoic and Mesoproterozoic eras. In SW Ireland, the sediments and volcanics had it easier, only being folded, thrusted and moderately metamorphosed, but the principle was the same: the newer Avalonia sediments and volcanics got plastered onto their respectively older rocks of Rosslare. A lot happened during this momentous meeting, including the intrusion of most of Ireland's great granite masses (the so-called batholiths of Leinster, Donegal, Galway and Newry).

'Ireland', in a sense, had the good fortune to mesh where the Iapetus Ocean had once been – in a join running diagonally from NE to SW across the middle of the country (Drogheda to Shannon). This is known today as the Iapetus Suture. Because it was two *continents* that collided, this suture in its entirety – and the general package of rocks on either side – actually stretches from northern Norway, through Britain and Ireland, and into North America right down through the Appalachian Mountains.

This Caledonian orogeny formed two very large continents on Earth: Gondwana, comprising modern India, Africa, South

America and others; and Laurussia, comprising modern North America, the Baltic regions and Russia. It is important to realise that 'Ireland' was not what we think of as a distinct entity: it was just an anonymous part of the larger landmass of Laurussia. And we, as part of this large landmass, were still drifting northwards, getting ever closer to the equator. During the Devonian Period, after the continents of Laurentia and Avalonia had amalgamated, deserts formed. We were covered in deep rusty-red sandstones, occasionally dotted with salty lakes; we had sand dunes; there were mountains that shed conglomerates as a result of infrequent, but heavy, rains. It was hot. We can best see these red sandstone rocks today, coloured by their oxidised rusty iron minerals, exposed in large tracts of counties Waterford, Cork and Kerry.

By the Carboniferous Period we were finally on the equator and having a well-earned bask. But Gondwana and Laurussia were still tectonically twitchy and there was ongoing splitting and twisting. Our deserts were subsequently enveloped in warm, tropical, shallow seas as we sank gently beneath the waves. But not too far beneath. The limestones that cover much of central Ireland, and that now form a sort of limey plaster cast covering the massive scar of the Iapetus Suture, were formed during the early to mid-Carboniferous Period. They are packed full of tropical fossil sea life. And we were lucky: while 'Ireland' was basking in Caribbean-style conditions, further to the south, Gondwana was having an ice age. But this idyll was not to last. Gondwana and Laurussia were themselves not that far apart and were coming together. And we were about to feel the effects. This time we were slightly more protected from the continental crash than during the previous Caledonian orogeny because the collision in question was to the south of us – Iberia and Africa (as part of Gondwana) were about to obliquely ram into Europe (as part of Laurussia) to produce what is known to European geologists as the Variscan orogeny.

The Variscan orogeny occurred at the end of the Carboniferous Period and continued into the Early Permian. The frictional drag on the underside of the continental 'European' plate by the subducting plate to the south caused us to get stretched a bit. Then, as the collision proper got underway, we were squashed. That's how it goes for land close to a subduction zone: first you get stretched and pulled apart, then you get crushed in a continent-sized thumbscrew. It was during the Variscan orogeny that the folding and thrusting of the red sandstones (those desert rocks from the Devonian) that today form the Munster mountains took place. We had now become part of the fifth and, to date, last of the great supercontinents: Pangaea (see Figures 1.5 and 1.6).

Ireland was still a kind of anonymous entity. It was just a part of the middle of a huge landmass that was continuing on its journey northwards. During the Permian Period, we were about 10 degrees north of the equator (see Figure 1.6) and we were back again into hot, dry desert conditions. More red sands dominated, although in contrast to the Devonian deserts, this time we were on the margin of a shallow, periodically evaporating sea as well, and that is why we have the gypsum of Kingscourt (County Cavan) and the rock salt of Antrim, both minerals being formed from evaporating seawater. The subsequent Triassic Period continued the desert theme. It is worth noting that onshore Ireland has today relatively few rocks dating from the Permian, Triassic, Jurassic or Cretaceous periods; offshore Ireland, by contrast, has plenty.

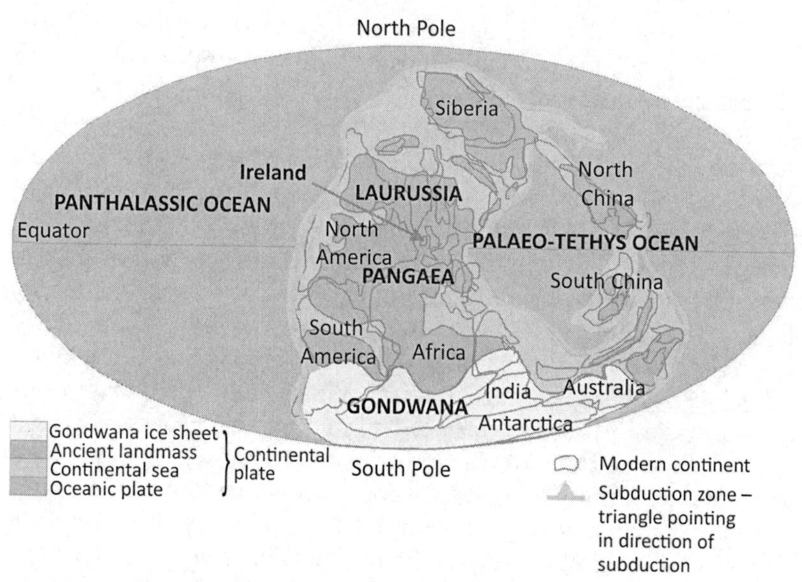

Figure 1.5. Map of the world during the Upper Carboniferous Period (Pennsylvanian) (318–299 million years ago).

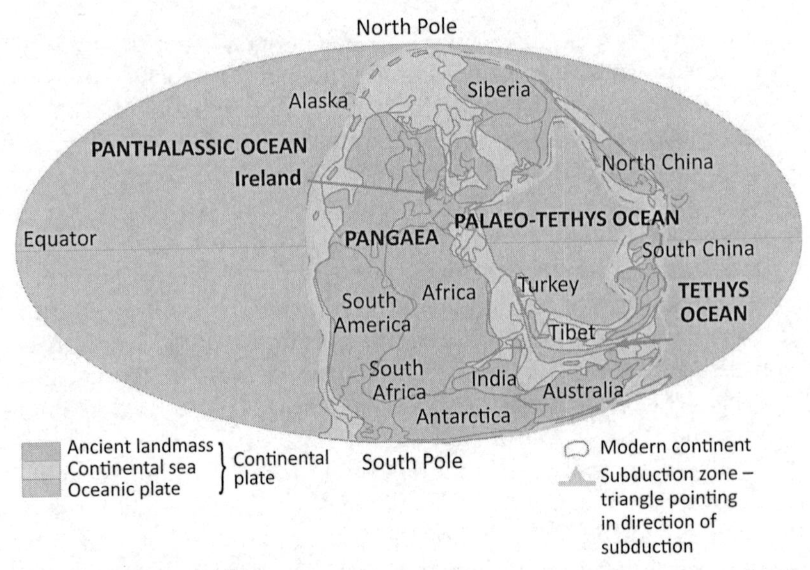

Figure 1.6. Map of the world during the Permian and Triassic periods (spanning 299–200 million years ago).

During the Jurassic Period, Ireland was some 30 degrees north of the equator and probably mostly land but with shallow seas covering our southern, eastern and northern margins. Did we have dinosaurs at this time? Who knows. Very few rocks of this period survive in Ireland today, although we have plenty of Jurassic sediments offshore. And why is that? Because the supercontinent of Pangaea was itself starting to split, and we found ourselves at the margins again – which, as already mentioned, is where things happen. A tearing phase began: some of the land sank (via faults, the blocks of rock acting like slowly downgoing lifts); sedimentary basins formed in what is now offshore Ireland, and these filled with Mesozoic sediments. This is the essence of the story of our own Jurassic and Cretaceous periods – it mostly happened in what is now preserved in our offshore waters (see Figure 1.7). But there are tantalising hints that, at least for a while, we (the current land of Ireland) might have been completely covered by the enigmatic Cretaceous sea that produced the famous white chalk that would later become the Cliffs of Dover (England) and the Cliffs of Cap Blanc Nez near Calais (France). Chalk exists under the Antrim

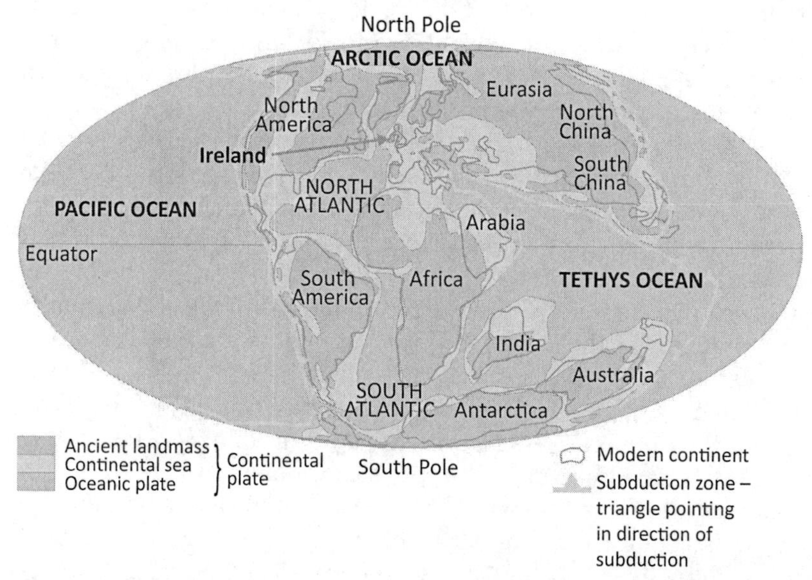

Figure 1.7. Map of the world during the Cretaceous Period (145–65 million years ago).

basalts in Northern Ireland and in small, and rare, patches in the Republic. Exactly why and how so much chalk was deposited in the world during the Cretaceous Period is still a mystery.

During the Cretaceous Period, while shallow chalk seas dominated locally, something was happening out to the southwest of us: the North Atlantic Ocean was starting to open via more splitting and tearing of the Pangaea supercontinent: its time was geologically up. And it continued to open well into the Paleogene Period – completely ignoring events such as the dinosaurs getting whacked by an Everest-sized meteorite at the end of the Cretaceous Period. The opening of the North Atlantic during the Paleogene Period was to have a major effect upon us, because, during this tearing event, entire areas of land were ripped apart down to the mantle, which caused locally cataclysmic volcanic activity to occur. The mantle's hot, red basalt blood flooded over the Antrim plateau, part of which is now the Giant's Causeway World Heritage Site. During this period, Ireland was about 45 degrees north of the equator, and the climate was hot and wet. Today, we can see that rocks that were exposed during the Paleogene show evidence of tropical weathering, which is odd because Ireland was not located near the tropics. We can thus infer that Earth's climate as a whole was significantly warmer than it is now.

Although the Lough Neagh Basin and its lignites (a type of soft coal) were formed during the Paleogene Period, much of the other evidence (rocks) from both the Paleogene and the Neogene periods is fragmentary. What we can say for sure, however, is that Ireland's older Carboniferous limestones were being eroded away at a prodigious rate (53 metres per million years is one estimate) and that during the Neogene Period our climate was warmer and less seasonal than it is today.

Which brings us to Ireland's landscape. The Neogene Period marks the point where scientists can start to identify what factors led to the development of the physical landscape that we see in Ireland today. Research on Ireland's landscape is a hot topic among geologists and geographers, with great effort being expended on unravelling exactly why we have the physical landscape features that we do have. For example, the development of the Wicklow Mountains (which were not always mountainous) got underway during the Neogene Period, and their rise seems to be one of several curious side effects of the North Atlantic opening.

After the Neogene Period came the Quaternary, and this means only one thing: ice. And warm times. And ice. And warm times. Yes, it alternated between ice ages and warmer times on many occasions over the last 2 million years. But it was the ice phases that dominated. At the end of what most people term 'the Ice Age', Ireland finally became recognisable as an island when a rise in global sea level occurred (due, in fact, to the cataclysmic draining into the North Atlantic of a burst gargantuan glacial lake in North America) that cut us off from the rest of Europe, including what is now Britain. Only at the very last, geologically speaking, are we recognisable as the island of Ireland.

It is worth noting that the popular term 'Ice Age' actually only refers to the last 'ice age', which ended some 10,000 years ago. The Quaternary Period was hugely important in terms of shaping today's Ireland and giving us our own living conditions because what happened then continues to profoundly influence our agricultural sector and our water system today, and, via glacial deposits, produced much of the aggregate we use for all kinds of building and road works. Despite its massive importance, the Quaternary Period is often disparagingly dismissed by many geologists as 'mere overburden' (or even 'the gardening'): it is usually left for geographers and engineers to grapple with. But it is good stuff, and a small and passionate band of geologists do devote their lives to it. Many elements of our landscape derive from the Quaternary: drumlins, eskers, kettle holes, raised beaches and much else besides. Plus, during the Quaternary we had giant deer, woolly mammoths, wolves and bears.

And finally, Ireland gets to 55 degrees north. Man arrives. And he makes changes of his own. This is where archaeology and history start to add to the mix. It is sobering to consider that much of the way Ireland looks in terms of trees (or the lack of them), as well as bogs and fields – essentially, what you see when you look down at Ireland from an aeroplane – is just about entirely man-made.

Ireland has an exceptionally rich geological history. Our rocks bear witness to many of the planet's greatest events. We were born in two halves, the two halves independently wandering about the face of the Earth as parts of various continents and supercontinents, we were geo-surgically sutured together and then had all kinds of wild adventures involving deserts, tropical seas, volcanic eruptions and ice ages. Since the Cambrian Period, we have continentally sailed northwards from the southern hemisphere, across the equator, and off into the higher latitudes of the northern hemisphere: the latter being a new experience for SE Ireland, but perhaps giving a strange sense of déjà vu for NW Ireland. There is absolutely no doubt. Ireland has had an extraordinary history.

2

THE IRISH FLAG IN A DIFFERENT LIGHT

Question: Do you recognise what is in the photograph in Plate Figure 1?

Answer: It is a unique version of the Irish flag. According to ancient Celtic legend (*The Annals of the Three Micas*), the green, white and orange (chlorite, muscovite and biotite) are the sept colours of Phil O'Silicate, traditional sheet-meter to Rí Roc na hÉireann (King Rock of Ireland).

This is a rare example of a natural version of a national flag found in a rock from that nation. In this case, it is the Irish flag and, as an added bonus, it has been formed from three members of the one mineral family – the micas. The caption includes some word play around Irish names and mineral terminology. The minerals chlorite, muscovite and biotite are all members of the mica family, and the micas are themselves members of the larger phyllosilicate mineral group: 'phyllosilicate' just means 'leaf-like silicate' in allusion to their all having sheet-like crystal structures and usually occurring as flakes. This mineral group can be thought of as akin to the concept of a clan or a sept for a community of related minerals. And the term 'phyllosilicate' quite naturally transforms into the Irish-sounding personal name of 'Phil O'Silicate'. It all fits together pleasingly well.

I discovered this 'flag', which is only some 1.5mm in length, in a very thin slice of Leinster Granite (expertly ground down to 0.03mm by Tom Culligan of University College Dublin) that I had collected from the Dublin Mountains. It could not be more Irish. This 'flag' was then photographed by attaching a camera to a petrological microscope, a type of microscope that geologists use to examine very thin slices of rocks and minerals, and by photographing the specimen under a special type of light that the microscope produces called polarised light.

Here is a fun project for geologists, mineral enthusiasts and flag fanatics: can you find and record other examples of national flags that are reproduced by naturally occurring minerals? To help get you started: simply turn the Irish flag the other way around so that it now becomes vertical orange, white and green stripes ... instantly, the flag of the Ivory Coast. However, one could *not* turn it sideways and claim to have the flag of India because there is no complex blue circular design within the white band. While complex flags, i.e. those that include writing and/or symbols, are not going to be reproduced by nature, there is a very good chance that many of the simpler flags could be found naturally if you know how to look.

Become a 'mineral vexillologist' (a newly invented hobby), and see how many mineral flags you can discover. Publish them online, or in an appropriate magazine.

3

The Blue Waterfall

Once in a blue moon, you come across a photograph that just stops you in your tracks. It demands a double take. What you see verges on the unbelievable. As far as I'm concerned, the photograph shown in Plate Figure 2 is in this class. And it is from County Waterford.

It shows an azure-blue and verdant-green stunningly beautiful crystalline 'waterfall'. It's the sort of thing that might grace the pages of *National Geographic* having been discovered by an expedition to some exotic location, perhaps hidden away from the world in a remote jungle cave. But it is not. It is one of the most amazing sights in (yes, *in*) Ireland. It was named by Ike Wilson in 2001 and subsequently mapped and photographed by geologist and cave enthusiast Martin Critchley a year later (see the *Journal of the Mining Heritage Trust of Ireland*, December 2002).

To put this blue waterfall in context, first let me take you on a virtual tour of where you'll find it: Tankardstown Mine, Bunmahon, on the coast of County Waterford. Not a big mine, Tankardstown was opened around 1850 by the Mining Company of Ireland to extract copper ore. The mine closed in 1875 and reopened in 1905, only to finally close shortly afterwards in 1907. Until the mid-1970s, the mine's main shaft, called Heron's Shaft, could still be accessed. But it became blocked after a burning stunt lorry was crashed down into it during the making of a film. To this day, the lorry remains embedded in the shaft.

One alternative route into the mine is via an entrance 5 metres up a cliff (requiring a ladder) from the base of the beach. Once inside the entrance, you are faced with an *Indiana Jones* set of obstacles and dangers. While the adit (tunnel) is passable, one must beware of knocking oneself out by clattering one's head on the irregular roof and walls – hard hats are a must. There is a section in the tunnel where, just as in *Indiana Jones*, the gaping black maw of a vertical shaft suddenly appears. This shaft can only be crossed by very carefully walking, tight-rope-balancing style, on two thin metal pipes that lie across it. Water in the tunnel can easily get to over 1 metre deep, and you cannot always see your footing.

Worse, in one of the watery sections of the tunnel there lurks another one of those sudden vertical mine shafts. But this second shaft descends precipitously into Hades and it can be near impossible to detect the danger because it has flooded right up to the level of water you are walking in. Miss your step here and you will simply sink into a dark, watery nothingness.

However, if you have the bravery and the skill to negotiate these terrifying hazards, after some 60 metres you will come to one of the most jaw-dropping sights in, on, or under Ireland – the Blue Waterfall. Fabulous blues and greens form a crystalline cascade of breathtaking hues over a vertical height of more than 6 metres (over 20 feet). Gazing up at this must be magical: alas, I have not personally seen it. The waterfall is made from something analogous to the type of flowstone that one commonly sees in limestone caves. The difference here is that under Tankardstown this extraordinary flowstone is not formed of common calcite but of a range of secondary copper minerals that have precipitated out after being leached from their original copper minerals in the local bedrock.

To describe the blue as 'azure' is perfectly accurate. Much of the 'blue waterfall' is made of the copper mineral azurite. There is also probably some langite (another copper mineral) and possibly others as yet unidentified. But one would not want to desecrate such a beautiful and delicate creation of nature by too much sampling. This is a wondrous symbiotic creation of nature and man.

NOTE: *There is no general right of access to abandoned mines in Ireland; all abandoned mines are in the ownership of the State, and access is* *controlled by the Minister for Communications, Energy and Natural Resources. Under no circumstances should you explore any abandoned mines except under expert guidance and with the full permission of all parties concerned.*

4

Crusty Seán Has a Lager in Kilkenny

When Mother Nature was handing out exceptionally preserved fossils, Kilkenny must have been near the head of the queue. Not only does Kilkenny host the extraordinary and geologically famous fossil plants of Kiltorcan, near Ballyhale, and the superb and rare fossil amphibians from the Jarrow Colliery coals of the Castlecomer Plateau, it also hides yet a third extraordinary fossil occurrence.

But first, the title of this chapter, which I had better explain. It is not about a neo-pagan hippie called Seán who wants to quench his thirst with a pint of Harp in the grounds of Kilkenny Castle. Rather, 'Crusty Seán' is a bit of nominal humour around breaking up the word 'crustacean' into an Irish-sounding name, while *Lager* is the German word for 'resting place' and is a shortened version of the term *Konservat-Lagerstätte*. A *Konservat-Lagerstätte* is the technical term in palaeontology for an occurrence of exceptionally well-preserved fossils, often displaying features never normally seen in the fossil record, such as soft body parts and/or internal organs. Putting it all together, the title refers to a truly remarkable occurrence of fossil arthropods recovered from fine-grained mud rocks in drill-core taken from the Upper Carboniferous (Pennsylvanian Epoch) of the Castlecomer Plateau. The fossil preservation is world-class, has had a short book devoted to it, and has been researched by Irish geologists, including Patrick Orr of University College Dublin. And, like many an Irish geo-wonder, 'Crusty Seán' has remained almost unknown to the general public. Well, no longer!

What is preserved here? Does our Crusty Seán have a proper name? For those who get a perverse kick out of complicated names, we are talking about exceptionally preserved spinicaudatan conchostracan branchiopod crustaceans. For those who prefer the simpler version – clam shrimps. There are two varieties here: *Limnestheria ardra,* named after the Ardra drill-core in which it was first found, and *Limnestheria gracilis,* which is also a new species and was found in the Hollypark drill-core. Ardra and Hollypark are areas in the Castlecomer Plateau.

Clam shrimps are a group of arthropod animals that live inside two shells; the shells form an outer carapace and the animals can superficially look like small molluscs. The fossil record of clam shrimps can be traced back to at least the Devonian Period. I'm not sure if the arch-Catholic and arch-conservative Éamon de Valera would have entirely approved of clam shrimps once living in 'Ireland': these Kilkenny clam shrimps are of the Limnadiidae family, and they have notorious sex lives. They can reproduce the normal way involving a male and a female; they can reproduce by the mating of two hermaphrodites; and they can reproduce by the union between a male and a hermaphrodite. Dev's ghost has just fainted at the very thought.

With all this reproductive activity to enjoy, a clam shrimp needs something to hang on with. And our Kilkenny lads (and other genders) have just the thing. One of the exceptionally preserved features in the fine, green, muddy rocks is a clear impression of claspers. And yes, claspers were used to hang on to a partner (of whatever persuasion) while the business of reproduction was being actively engaged in.

As well as claspers, other things not normally seen in the fossil record but clearly visible in the Kilkenny muds are antennae, mandibles (jaws) and telsons (tails), all preserved in excellent detail, often in full 3D (normally fossils are squashed into flat 2D). The morphology we can see in the fossil clam shrimps of Kilkenny allows us to gauge how much the group has evolved between the Upper Carboniferous (Pennsylvanian) and today. As it turns out – not much. Evidently, their 'what the hell' reproductive strategy, among other factors, has allowed them to successfully survive all manner of mass extinction events and other geological travails over the past 300 million years. Makes you think!

5

FIFTY SHADES OF GREY – LOUGHSHINNY REVEALED

Beds … cleavage … thrusting … veined members! It's all been filmed at Loughshinny. People the world over have been stunned by what they have witnessed.

What kind of north Dublin rural debauchery is this? Has the Hell Fire Club reformed and overrun the seaside village of Loughshinny? There's no need for moral outrage: it's Mother Nature's own beachfront exhibition of contorted rocky positions. Loughshinny enjoys a special place in Irish geology because of the spectacular way that some huge tectonic events have impacted on it over the past 325 million years. So, what is going on?

Start at Loughshinny beach car park, walk due south 100 metres along the beach and look at the rocks in the cliff in front of you. You cannot miss them (see Plate Figure 3). What you are seeing are structures that are famous – textbook-quality folding of the rock layers. The rock is a mix of thicker, light-grey limestones and thinner, dark-grey, limey mudstones. In detail, there are at least 50 shades of grey on display. And the grey here is geologically sexy. The layers have been tectonically crumpled in a very special way.

Some 325 million years ago, during the early Carboniferous Period (Mississippian Epoch), Ireland was on the equator. We were part of the southern margin of a very large continent called Laurussia, which comprised most of what we now call North America and Russia. We basked in warm tropical seas, jam-packed with wonderful life, all doing its thing: teeming and thriving

– somewhat like the Great Barrier Reef or the Bahamas today. However, trouble was brewing to the south. Spain and Africa, which were in the southern hemisphere, were irrevocably heading northwards on a collision course with us. The stage was being set for 'Ireland', and what would become northern and central Europe, to be involved in one of those great slow-motion continental crashes that happen every geologically so often. This particular crash of continents – and there have been many over the aeons – is known as the Variscan orogeny.

The odd thing about these types of huge collisions is that before they squash you they first pull you apart a bit. You get somewhat extended before you get mashed. The extension phase of the Variscan orogeny is very important in Ireland. One of its effects was to stretch the pre-existing rocks (in the Dublin region these were the metamorphosed sedimentary and volcanic rocks from the earlier Ordovician and Silurian periods), so that parts of Ireland were slowly pulled apart, causing the local development of ever-deepening sedimentary basins to form because parts of the land were sinking. One of these basins was the Dublin Basin, which on its east side stretches from Skerries, in the north, to Blackrock village, in the south, but widens westwards to reach as far as counties Westmeath and Offaly. Think of it as something like today's East African Rift, but under a kilometre or two of warm water.

Loughshinny lay very close to the northern margin of the Dublin Basin; Lucan, in County Dublin, was near the centre of it, on its eastern side. As this basin started to open and deepen, steep shelves (like mini continental shelves) developed off the shallow-water reef limestones at its edges. Frequent earthquakes and tropical storms sent cascades of fragmented limestones pouring down the shelf slopes into the basin itself, settling out as coarser grains first and finest grains last, under the deeper water. These undersea cascades of sediment produced a rock type called a 'turbidite', which is a rock formed from undersea turbidity currents, i.e. a fast-flowing current of sediment-laden water. We can see these turbidites today as the thicker limestone beds exposed at Loughshinny. In between these limestone turbidites are the thinner, darker-grey shales. The shales represent the periods between the thick cascading turbidite influxes when the background sedimentation was quite slow, resulting in thin accumulations of fine-grained, limey clays containing very dark-grey organic remains.

However, there came a time during the Variscan orogeny when the crustal stretching, with its widening and deepening of the Dublin Basin, stopped and the slow, contractive mashing period started. Spain and Africa collided full-on with what is now Central Europe, and to which Ireland was attached (the collision was slightly oblique to us), resulting in some of the mountains in Europe that we can still see today. Although 'Ireland' was hundreds of kilometres from the main collision, we still felt its effects – most obviously in the development of the Cork and Kerry mountains, where the red sandstones of the Devonian Period were crumpled and thrust over themselves. But here's the thing: while the Variscan orogeny caused folding effects across the Midlands and possibly as far north as Northern Ireland, these effects were not very pronounced. Often we just got some gentle undulations in the Carboniferous limestones. So why the anomalously spectacular display of folding at Loughshinny?

The reasons are threefold. First, it is the nature of the rocks themselves. They are not pure limestones, which would fold and buckle a little, but after a point would just fracture and break and so be deformed mainly by various types of faults cutting relatively gentle folds (think of the Burren, for example). They are also not pure limey shales and muds, which would slip and slide and deform more like a platy plasticine. The Loughshinny rocks are, in fact, a very particular sandwich of alternating beds of shales (soft and slippy when deforming) and limestones (hard, and prone to cracking after a certain point). This combination of alternating rock types with very different responses to tectonic stress (the squashing from the Variscan orogeny) are why the folds in Loughshinny are the way they are. The limestones are trying to remain 'solid', while the shales are literally going with the flow, and the competing physical properties of the two rocks produce a compromise type of structure: a zigzag (chevron) style of folding.

Second, we must remember that the Loughshinny rocks were formed at/near the northern margin of the Dublin Basin, where there are many faults – the same ones that were active and allowed the basin to open in the first place during the stretching phase.

When the subsequent squashing phase started, most of the deformation got taken up by these pre-existing structures and the rocks associated with them. As a rule, rocks on the margins of a sedimentary basin, like the Dublin Basin, will deform more than the rocks in the central parts. Loughshinny is at the margins so one would naturally expect more deformation than at Lucan. Yet this does not seem to entirely explain the really very intense folding at Loughshinny, which was far north of the Variscan collision front itself. One must be aware that this type of folding is *not* seen in comparable rocks to the west and elsewhere. There must be another factor at play.

The third factor is that to the south lies the very large and very hard Leinster Granite, which had formed during the Devonian Period, some one hundred million years earlier. This huge lump of granite seems to have acted like a rigid block during the Variscan orogeny such that most of the stress that was put upon it at its southern end (modern Wexford) was pretty much wholly transferred to its northern end (Blackrock in south County Dublin), which, in turn, directly affected all the rocks at the margins of the Dublin Basin on the basin's overall eastern side. To better understand what I mean, consider the analogy of a rectangular block of wood with some plasticine at one end: if you give the block a push from the other end, all the pressure is directly transferred to the plasticine and will fold and squish it, while the block of wood itself will remain undeformed, acting merely to transfer the pressure. For Loughshinny, the granite is the wood, the limestones are the plasticine, and the push is from Iberia.

Thus, in a nutshell, the Loughshinny folds came about by a direct transfer of stress northwards, via the Leinster Granite (the whole Leinster block being termed a 'massif'), as a result of the Variscan orogeny acting upon the special mix of alternating limestones and shales that occurred at the northern margin of the Dublin Basin.

The northerly directed pressure from Spain/Africa shunting us from the south also meant that the clay minerals in the shales became quite well aligned. When clays in a sedimentary rock are subjected to tectonic pressure, they have a habit of dissolving and reforming in an orientation that minimises the stress being put upon them. When this happens, we see a planar structure developed, which tends to line up not with the original bedding, because this can be contorted in many different directions, but at right angles to the direction of the maximum pressure. In the Loughshinny case, the pushing was from south to north, so the clay minerals lined up east to west – it was their way of trying to minimise the stress being put upon them. These planes, defined by the aligned clays, pass though the shales and, because they can cause the shales to split and cleave in a direction other than their original bedding planes, they are called cleavage planes. The consequence of all this is the cleavage one sees in Loughshinny.

Do we also have 50 shades of white? When you visit Loughshinny, don't forget to look at the numerous white veins in the limestone members (i.e. subparts) of the rock units. These too are directly related to the deformation that went on during the Variscan orogeny. What happened was that the internal parts of the thick, grey, brittle limestones fractured, causing holes and veins to open and to be filled with soft, white calcite. These holes/cracks are often curved in a distinctive way, called 'en echelon'. If you look at them and consider which way they must have opened in order to fill with the calcite, you can deduce that they almost all opened in the same direction – with the right-hand side of any given crack moving both down and off a bit to the right. It is the solid limestone's response to the stress: being moderately slowly pulled apart and locally cracking as part of the folding and deformation process. And this is the same sort of principle that causes echelons to be formed in bicycle races when there is a crosswind. It's all a response to stress.

As pointed out earlier, the folds at Loughshinny are world-famous. They have been the subject of countless photographs and many discussions in national and international books and articles. If you are lucky enough to live in or near Loughshinny, you have a lot to be proud of.

6

When Is Not-a-Volcano a Volcano? When It's a Sugar Loaf

Scherzo: The Great Sugar Loaf of County Wicklow

Was ever a mountain so misunderstood as the Great Sugar Loaf, south of Bray, County Wicklow? This conical mountain has achieved a fame and infamy out of all proportion to its 501-metre height.

The Great Sugar Loaf (see Plate Figure 4) stands guard over the Wicklow village of Kilmacanogue like a brooding Irish god. It dominates the landscape because it has no nearby competitor and also because it has a very distinctive shape. In this latter respect (only) it can be compared to the Matterhorn on the Swiss–Italian border, which, because of its shape and isolated context, looks bigger than it actually is.

According to some of the older generation of Dubliners I have talked to, they allegedly learned in primary and secondary school – between the 1940s and the 1980s – that the Sugar Loaf was an extinct volcano. Or, more scary still, that it wasn't even extinct but merely *dormant*. More recently, based on a perusal of some internet forum comments, some members of the public used to believe – and a few still do believe – that this mountain is made of granite and is some 200 million years old. Others consider that its origins go back not millions but billions of years, and that it was once the size of an Alp but has reduced in size considerably over (only) the last few centuries. There is a great deal of confusion and misunderstanding out there.

The public debate is complicated further by the fact that there are several 'Sugar Loaf mountains' in Ireland. Plus there is a Sugar Loaf Mountain in Wales and a famous one in Rio de Janeiro in Brazil. These are all geologically dissimilar and add even more layers of confusion to explaining the Great Sugar Loaf in Wicklow, with many different stories getting interchanged and becoming garbled.

But one cannot entirely blame the public. Remember the old saying, 'If it looks like a duck, waddles like a duck, and quacks like a duck, then it probably is a duck'? Well, this isn't true in the case of the Great Sugar Loaf. It does look like a small volcano; it does have the feel of a volcano; and many claim they were once told *by people in authority* that it *is* a volcano.

Before we go any further: the Sugar Loaf is *not* a volcano! It never was a volcano, and it never will be. It has nothing to do with volcanoes. End of.

What, then, is it? And why does it look the way it does?

The Great Sugar Loaf is, for the most part, made of metamorphic whitish to pinkish quartzite, i.e. relatively pure quartz sand grains welded together with a quartz cement as a result of an older, original, pure sand being metamorphosed during the Caledonian orogeny (mountain-building event) some 400 million years ago (see Chapter 1). The original rock, before it was metamorphosed, was a relatively pure quartz sandstone that had formed some 500 million years ago during the Cambrian Period. This part of the answer is straightforward enough. But it doesn't answer the question of why the sandstone of the original was so pure, or why the metamorphic version of the rock now looks like a volcanic cone.

The sedimentary environment for the original, Cambrian Period, pure sandstones of the Great Sugar Loaf is difficult to determine. When the sandstones were being laid down, Ireland was still in two separate halves, with an ocean in between (see Chapter 1). The sands of the Sugar Loaf were being deposited, underwater, on the northern, oceanward side of a micro-continent called Avalonia, which is where the southern half of Ireland comes from. The rocks that surround these pure sandstones are turbidites (layered rocks formed from sediments sweeping down off a continental shelf and settling in deeper waters) and mudstones. These turbidites and mudstones were probably laid down in an ocean basin adjacent to the northern Avalonia continental shelf. Modern-day equivalent environments might be offshore Newfoundland or the Congo.

Recent research by Professor Peter Haughton of University College Dublin (see also Chapter 14) suggests that the original pure quartz sands arrived into this ocean basin from higher up on the original (Avalonia) continental shelf, the idea being that these pure sands slid into the deeper water basin en masse, possibly as a result of being juddered loose by large earthquakes. This would mean they originally formed close to Avalonia's ancient shore. The original sands might have been deposited in a delta from rivers running north off the land of Avalonia, and maybe they represent the sandy, eroded remains of a quartz-rich igneous or very metamorphosed suite of rocks. This is currently the best guess.

So, why the conical shape? As we have seen above, the original sandstones formed 500 million years ago during the Cambrian Period; these sandstones were metamorphosed into quartzites 400 million years ago during the Devonian Period as a result of the Avalonia micro-continent running into the continent of Laurentia: this was the Caledonian orogeny. Subsequently, they remained buried under a variety of later Mesozoic and Cenozoic rocks for the best part of the next 380 million years. They finally seemed to have reached the surface and were subjected to erosion only in the last 20 million years, or possibly even less.

This last phase is crucially important and was the result of the whole of Leinster being uplifted during the Cenozoic Era as one of the consequences of the opening of the North Atlantic Ocean. The quartzites are quite pure and hard, the whole mass being relatively uniform in composition and structure. Therefore, when weathering and erosion processes act on this, it is akin to trying to erode a mountain of almost pure quartz. Quartz is highly resistant to erosion and, because it is both hard and homogeneous, will erode equally in all exposed directions.

Surrounding the hard quartzites is a softer rock called schist (the now-metamorphosed Ordovician turbidites and muds), which will erode faster than the quartzite, thereby exposing more of the quartzite at the base as it does so. As more quartzite is exposed, a conical shape develops (due to how the quartzite itself erodes), which becomes more pronounced with time. A little like one's gums receding and exposing more of a tooth.

Note that the nearby Dublin and Wicklow granite mountains also contain quartz, but not as much. The granites do not erode conically but form more rounded, softer erosion profiles. This is because they contain, in addition to hard quartz, more feldspar in their make-up, and feldspar is slightly softer than quartz, erodes faster, and allows for a more rounded shape to develop.

The conical shape of the Great Sugar Loaf, therefore, has a completely different origin to that of a volcano (see Plate Figure 4). In the case of a volcano – specifically a stratovolcano, which has the classic conical profile (e.g. Mount Fuji, Mount Pinatubo, Mount St Helens and Mount Rainier) – the shape is the result of lava and volcanic ejecta being built up, eruption after eruption, so that a critical angle is reached for the built-up pile, where the material comes to rest before sliding down. It's like building up any unconstrained pile of loose material: the result is always some sort of cone shape. By contrast, the Great Sugar Loaf gets its shape from the erosion and weathering of very resistant quartzite. And that same principle applies to other quartzite mountains in Ireland, such as Croagh Patrick (County Mayo), the Twelve Bens (Connemara) and Errigal (County Donegal).

Just for the record: Wicklow's Little Sugar Loaf is not volcanic, either. It is also Cambrian quartzite. The Sugar Loaf Mountain in West Cork, southwest of Glengariff, has also been described as an old Irish volcano. It is not. It is a conically eroded pile of Devonian sandstones. The West Cork Sugar Loaf is a kind of mineralogical cousin of the Great Sugar Loaf, i.e. mostly quartz, but from rocks that are some 150 million years younger and softer than the quartzites of the Great Sugar Loaf, and that were formed in a very different original sedimentary environment: a desert environment with flash flooding, as opposed to turbites under the sea. Furthermore, the West Cork 'Loaf has only been 'lightly baked' when compared with the older Cambrian quartzites, this time due to the younger Variscan orogeny, not the older Caledonian orogeny. Sugarloaf Hill in the Knockmealdown Mountains of Waterford is not an old volcano, either. And Sugarloaf Mountain in West Wicklow is actually made of Ordovician schists. Thus, not one of the Irish Sugar Loaf mountains is volcanic. Nor is the Sugar Loaf in Brazil, nor that in Wales.

If you want a genuinely volcanic 'sugar loaf' edifice, then you have to go to the Sugar Loaf Islands situated off the west coast of New Plymouth, now part of the Taranaki volcanic park, on the North Island of New Zealand. These islands *are* genuinely volcanic. But further away from Kilmacanogue you could hardly get.

Coda: What Is a Sugar Loaf Anyway?

A sugar loaf is a conical 'loaf' of refined white sugar, with origins that can be traced at least as far back as twelfth-century Jordan. At different times and from different manufacturers, a sugar loaf could weigh between 3 and 35 pounds: the lower the weight, the finer the grade. The shape results from the method of refining the raw sugar: after a series of boiling and filtering processes, the sugar liquid is poured into an inverted conical mould with a hole at the bottom through which could drain any remaining molasses, syrup or other non-crystalline material. When the remaining white sugar was pure and hard, the conical mould could be removed, and the now-conical sugar loaf itself would be set upright on its flat base. It would then be trimmed and packaged for sale. To get sugar off such a loaf required special strong metal pliers called 'sugar nips'.

And because sugar, for most of its history, was expensive, the quantities nipped off and used in drinks and desserts were small. One sugar loaf could last for years. If you are in Dublin, you can go and see a real sugar loaf on display in the kitchen pantry in the Georgian House Museum at 29 Fitzwilliam Street Lower.

7

Ireland's 'Jurassic Park' – Closed Due to Flooding

Irish Dinosaurs

Dinosaurs started to evolve during the Triassic Period and really came into their own during the following Jurassic and Cretaceous periods. Unfortunately for Irish children (and grown-ups), the Republic of Ireland does not have any dinosaurs, and there is very little prospect of ever finding any. No kentrosaurs in Kerry, no tyrannosaurs in Tyrellspass, no oviraptors in Omey, and certainly no parasaurolophuses in Pallaskenry.

The Republic not only has very few rocks of the right age, but the very few rocks of the right age that we do have are of the wrong type. We have been doubly cursed. Almost without exception, all our extant dinosaur-age rocks were originally laid down underwater. But dinosaurs did not live underwater: they were all landlubbers. And creatures such as ichthyosaurs and plesiosaurs were marine reptiles, not dinosaurs – so they don't count.

However … the island of Ireland, as opposed to the Republic of Ireland, *does* have dinosaurs. A grand total of two (see Plate Figure 5). These are represented by two bones, one each from a different species. Both bones are fragments of hind legs and both are from dinosaurs of the Lower, possibly Middle, Jurassic Period, about 200 million years ago. Both were found in Northern Ireland, and both were found over a twenty-year period of searching by the same man:

Roger Byrne, a school teacher and very dedicated amateur palaeontologist. Not only that, but the bones were also found in the same place: the beach at the Gobbins, Islandmagee, County Antrim. And these astonishing discoveries are made more impressive still because the fossil dinosaur bones themselves are black in a dark mudstone, and Roger Byrne found them against a backdrop of countless black cobbles of basalt. I wish I had his eyes. Sadly, Roger has now passed away, but he has already gone down in history.

So, what are our two species of dinosaur?

1. *Scelidosaurus*. The bone is a small 6cm fragment of a femur (upper leg bone). Scelidosaurs are a type of primitive armoured herbivorous dinosaur from the early Jurassic Period; they were about 4 metres (13 feet) long and characterised by having a body armour embedded in the skin of their backs, sides and tail that was made of bony plates covered in horn. They are also thought to be ancestors of the more famous ankylosaur group. It was a *Scelidosaurus* from Dorset in southern England that was the first almost complete dinosaur skeleton to be described (by Richard Owen in 1859, although not published until 1861), and, appropriately enough, considering our bone fragment, the name itself means 'leg lizard': this alludes to the fact that its hind legs were longer and larger than its front legs.

2. *Megalosaurus*. The bone is a small 10cm fragment of its tibia (lower leg bone) (see Plate Figure 5). *Megalosaurus* was among the first three species named as 'dinosaurian' by Richard Owen in 1841, although the name *Megalosaurus* itself had appeared several decades before. Despite its formative place in dinosaur history, true megalosaurs are still not that well characterised. 'Megalosaur' means 'big lizard'; it was a bipedal, early-Middle Jurassic Period meat-eating dinosaur about 9 metres (30 feet) long, belonging to the therapod family. It has also been found in England, France and (maybe) Portugal. In its day, *Megalosaurus* was an apex predator and could quite possibly have eaten scelidosaurs from time to time.

The question is: what are these two terrestrial dinosaurs doing in marine Antrim mudstones? The answer seems to be that they died on land, or in a river, close to shore and their dead bodies were washed out to sea, where they sank, decomposed, got covered in marine (or near-shore) sands and muds, and became fossilised. This type of process has been inferred for several other dinosaur occurrences in marine rocks elsewhere in the world. And, of course, we see analogous events today when large land animals get washed out to sea by flash floods or by a river in torrent, and so on. But our two dinosaurs were certainly not in their native habitat when they reached their final place of repose.

Ireland during the time of the dinosaurs was, at least in part, under water. During the Jurassic there was possibly more land exposed than during the Cretaceous, but we were often flooded by shallow seas. The real land-based 'Jurassic Park' was taking place not so very far from us in what is now southern England on one side and North America (which was geographically close to us at the time) on the other. It just Sod's Law that we (Ireland generally) were quite often the shallow wet bit in between. However, if we did once have the right rocks of the right age (which would have been Jurassic if anything), and there were dinosaur fossils within them, they were very comprehensively eroded away later by quasi-tropical weather during the Cenozoic Era.

On the plus side, a few fossils of marine reptiles (ichthyosaurs and plesiosaurs) have been found in Northern Ireland. But it doesn't quite make up for us not having the drama of real dinosaurs once thundering across our land – and ripping the heads off each other.

Side note: in one of the few places on the island of Ireland where we do have the right rocks (terrestrial sands and muds of ancient inland lakes, represented by the Sherwood Sandstone Group, which is found around Belfast) and of the right age (Triassic Period, when the dinosaurs started to evolve), the only body fossils that anyone has found are the remains of a few … tiny … water fleas: *Euestheria minuta*.

Two Irish Dino'ddities

Ever fancied tea with a *T. rex*? Meet the mega meat eater at the Dino Cafe in Castletownroche, near Mallow, County Cork. The Dino Cafe is a unique comestible emporium that boasts life-sized models of several species of dinosaur, as well as being the only permanent display of dinosaurs in Ireland. It is guaranteed to excite or terrify, depending on your age and nervous disposition. Kids love it! Tucking into your 'full Irish' with one of nature's greatest carnivores almost breathing down your neck can make you feel like a small, curly rasher. Your visit will certainly be memorable.

The creationist website christiananswers.net also has a little something to say about dinosaurs and Ireland. Let me quote:

> An Irish writer [around 900 AD] recorded an encounter with a large beast with 'iron' on its tail, which pointed backwards. Its head was shaped a little like a horse's. And it had thick legs with strong claws. These details match features of dinosaurs like the Kentrosaurus (KEN-tro-SOR-us) and Stegosaurus (STEG-oh-SOR-us). They had sharp-pointed spines on their tails, thick legs, strong claws and long skulls.

It is unlikely in the extreme that spiky-tailed dinosaurs were wandering around early medieval Ireland being documented by monks. We do know that Irish monks had invented whiskey around this time, but even if there had been some overzealous tasting from their new distillation process, they would merely have been seeing double, not seeing dinosaurs.

Unless ... could it really be? That Ireland in 900 AD should properly be called the land of Saints and Scholars and Stegosauruses?

8

The Aliens Have Landed!

Close Encounters

We have been visited. By things not of this world. Eight times.

Ireland has had close encounters of the third(ish) kind. But not by little green men, however appropriate that would have been. In one case, the alien smashed a house; in another, the alien smashed some potatoes (sound familiar?); and in a third, the alien allegedly killed a man. Violent little aliens!

These aliens are, of course, meteorites. Stones that are older than the Earth; stones that randomly fall out of the sky; stones that mostly derive from the asteroid belt, in a part of the solar system far, far away. Since 1750, there have been eight observed meteor falls, with meteorites recovered from seven. From these, we have captured the aliens: we have performed difficult alien autopsies, and we now keep them in our own versions of 'Area 51'. To put it scientifically, we have (for the most part) made sections of them, examined them by every means, found out their component minerals, and chemically and isotopically analysed them. And we have (for the most part) kept them safe in our museums and academic institutions.

METEORITE CLASSIFICATION

Not all meteorites are the same. Indeed, the meteorites collected from Ireland are of several different types. Here are some names you need to know.

- *Chondrite*: A very common variety of stony meteorite characterised by having a large percentage of its mass made of relatively pale-coloured, millimetre-sized chondrules. Chondrules are small, almost spherical blobs of silicate minerals (olivine and pyroxene being the commonest) that had originally melted in space before accreting in a new host (very alien-sounding).
- *H Chondrite*: A variety of chondrite that is characterised by having high levels of iron (the 'H' designation), and chondrules that are on the small side (about 0.3mm).
- *H5 Chondrite*: A type of H chondrite that has 'enjoyed' being metamorphosed to the degree that the chemical composition of the olivines and pyroxenes has been homogenised. A H5 chondrite contains some secondary minerals (i.e. minerals not part of the original make-up), and has blurry outlines to the chondrules.
- *H6 Chondrite*: A type of H chondrite that has 'enjoyed' being metamorphosed to the degree that the chemical composition of all the minerals has been homogenised. Chondrule outlines have been obliterated and no melting has occurred.
- *L Chondrite*: A variety of chondrite that is characterised by having low levels of iron (the 'L' designation) and chondrules that are on the medium side (about 0.7mm).
- *L3 Chondrite*: A type of L chondrite with almost no metamorphic features and sharp chondrules.
- *L6 Chondrite*: A type of L chondrite that has been metamorphosed to the degree that the chemical composition of all the minerals has been homogenised, chondrule outlines have been obliterated and no melting has occurred.
- *Stone uncl.*: This refers to a stony meteorite that has not (yet) been classified.

Ireland's Eight Alien Visitors

In the order that they descended from above …

1779 – Pettiswood, County Westmeath (53° 32'N, 7° 20'W)

An account of this fall was reported seventeen years after the event by publisher and businessman William Bingley in the *Gentleman's Magazine* (1796, Vol. 66). Bingley, at the time he wrote his article, was living in England, but at the time of the meteorite fall he had been living in Pettiswood in Westmeath. He apparently hadn't mentioned it earlier for fear of being ridiculed. As this is the first such account for Ireland, a few quotes are in order:

> Two pieces … have been in my possession since the year 1779, which actually descended, in a loud-peal of thunder, upon a meadow, situated at Pettiswood, co. Westmeath, in the kingdom of Ireland. The size and form of this cake … is that of a twopenny heart-cake, supposing all the parts were together. The two pieces of the cake I am describing weigh three ounces and a half, and, I suppose, form two thirds of the whole. Be the composition of this stone what else it may, it has been adjudged to be neither fossil, pyrite, nor petrifaction; … in short, it is not any mineral substance, nor is it similar to any stone known in the country.

A later part of Bingley's description reads like an extract from Somerville and Ross's *An Irish RM*:

> At the instant this rude lump descended, our little village was enveloped with the fumes of sulphur, which continued about six minutes. To its descent five witnesses are now living; three of whom reside in London. It lighted upon the wooden part of a harness, called a stradle, belonging to a filly drawing manure to a meadow, and broke into three pieces. At the same instant the affrighted beast fell to the earth, under her load; as did the two equally affrighted gassoons [boys], the drivers, who, in good Irish, came crying to me, with two pieces of the stone, declaring that themselves and the filly were all murdered by

> this thunderbolt; none of whom, however, received the least injury. The two pieces, when I received them, after the resurrection of the boys, were warm as milk just from the cow; whence it may naturally be concluded, that the cake came from a scorching atmosphere, and pretty well accounts for the outside of it, in its formation, and during its stay there, having been tinged to a whitish brown, whereas internally it is of a silver white, exactly like the materials whence it originated, supposing my conjecture a fact.

The Pettiswood meteorite is recorded in the catalogues of the Natural History Museum, London, but it is unclear if the stone itself is still there. Geologically, it remains 'Stone uncl.', i.e. unclassified.

1810 – Mooresfort, County Tipperary (52° 27'N, 8° 20'W)

The 12th of August 1810 was a dark and sultry day. Between 11.00 a.m. and 12.00 p.m. at the house of Maurice Crosby Moore, Mooresfort, County Tipperary, some workmen were laying lead in Mr Moore's gutters. Mr Moore himself had just started on his way to Limerick on business. But very soon after he left, the workmen were astonished to hear a peculiar whistling noise coming from the air. One of the men immediately thought that the chimney had caught fire; another that there was a swarm of bees approaching. Let Mr Moore himself, having returned to his house a few days later, take up the story (from a letter dated 22 May 1811 to Professor William Higgins):

> On looking up, they [the workmen] observed a small black cloud very low, carried by a different current of air from the mass of clouds, from whence they imagined this stone to have proceeded: it flew with the greatest velocity over their heads, and fell in a field about three hundred yards from the houses they saw it fall. It was immediately dug up, and taken into the steward's office, where it remained two hours cooling before it could be handled. This account I have had from many who were present, and agree in the one story. I saw myself the hole the stone made in the ground; it was not more than a foot in depth.

It is very doubtful that the stone was hot and had to be left for two hours to cool. The outer part of a meteorite does, of course, get hot enough to melt in the upper atmosphere, but it travels so fast that there is no time for the bulk of the stone to warm up – internally it's still mostly at chilly outer-space temperatures. Any hot bits tend to spall off. When it hits the ground, it will actually be quite cool and can be picked up easily.

More recently, Brian Mason and Hugo Wiik published a paper in the *American Museum Novitates* in which they describe a fragment of the Mooresfort meteorite:

> In hand specimen a broken surface of the Mooresfort meteorite is medium gray in color, with numerous rusty brown spots. Examination with a hand lens shows many chondrules (usually lighter in color than the groundmass) and occasional bronze grains of troilite. The specimen is seamed with extremely thin black veinlets. On a cut surface, abundant grains of nickel-iron are visible; these give rise to the rusty brown spots mentioned above. The meteorite is quite hard and not friable.

The Mooresfort meteorite had a total recovered mass of 3,520g and was classified as a H5 chondrite.

1813 – Patrickswell (Brasky), County Limerick (52° 34'N, 8° 47'W)

This meteorite hit the ground in various-sized fragments on 13 September 1813, between the towns of Adare and Patrickswell, County Limerick. The largest fragment from this fall is known as the Brasky Boulder, and it weighs about 27,000g. In total, some 50,000g (or about 65lb) of meteoritic material was recovered, and this makes it by far the largest single meteorite fall in historic times in either Ireland or Britain. The Brasky Boulder is now owned by the National Museum of Ireland.

Although this fall happened over 200 years ago, there are some interesting newspaper reports on what the locals experienced. Here is a clipping from the *Limerick Chronicle* of 11 September 1813:

> Yesterday morning, about 9 o'clock, there was a most dreadful thunder heard in the direction from Patrickswell, towards Adare and Rothkeale [*sic*] in this county. The peals were very violent and continued for a considerable time, and were accompanied by some awful appearances; large fragments of atmospheric stones.

Another report, quoted in *The Irish Times* (12 September 2011), specifically described the fall of the Brasky Boulder:

> One more very large mass passed with great rapidity and considerable noise at a small distance from me; it came to the ground on the lands of Brasky, and penetrated a very hard and dry earth about 2 feet. This was not taken up for two days; it appeared to be fractured in many places, and weighed about 65 lb!

And, of course, there's always one contrarian: here is a man writing to the *Limerick Evening Post* who blames it on the weather:

> Such a notion [of rocks from space] would involve the possibility of a total derangement by trivial causes of the steady and magnificent order of the planetary system. Monstrous absurdity! Daily experiment proves … the passage through them of the electric spark [lightning in the atmosphere], which would accelerate their chemical action in forming the concretions.

Geologically, the Limerick meteorite is a stony meteorite known as a H5 chondrite. H chondrites are among the more common variety of meteorite and they are thought to have originated in the asteroid belt (between Mars and Jupiter). We can reasonably speculate that some random collision between two asteroids in the belt millions of years ago sent debris flying in all directions, and that one of those pieces of debris happened to end up on a collision course with Limerick.

The official name of this meteorite is the Limerick meteorite; but it is also unofficially known as the Brasky Boulder, the Adare meteorite, and the Faha meteorite. Pieces of it are in national collections, notably Irish and British, and in private collections.

1844 – Killeter, County Tyrone (54° 40'N, 7° 40'W)

On 29 April 1844, several people witnessed a meteor shower at Killeter, near Castlederg, in West Tyrone. Only one moderately substantial piece was found, although there are reports of other tiny fragments.

It was Professor Reverend Samuel Haughton of Trinity College Dublin who in the 1867 *Proceedings of the Royal Irish Academy* first described the fall, including the locals' reactions, and giving some very early chemical analyses. He notes a report from a local that the biggest fragment was 'about as long as a joint of a little finger'; that the fall took place at about three or four o'clock on a sunny evening, but that three gentlemen on a nearby hill said there were clouds observed over Killeter (as well as the distant sounds of a German marching band); that the three gentlemen were told, when they came down, of a wonderful shower of stones that had taken place, which had spread over several fields; and that the locals had a remarkable lack of curiosity about the stones. This paper also reminds us that the name given to such stones in those days was not meteorites but 'aeroliths' (atmospheric stones), though both terms are closely related.

The total recovered mass of this fall was only 142g. The rock has been classified as a H6 chondrite.

1865 – Dundrum, County Tipperary (52° 33'N, 8° 2'W)

The second meteorite to hit Tipperary in less than 60 years and, to cap it all, the second one to hit on 12 August. What are the chances of that? Here's an eyewitness account published in the 1867 *Proceedings of the Royal Irish Academy*:

> I, John Johnson of the parish of Clonoulty near Cashel county Tipperary, was walking across my potato garden at the back of my house in company with Michael Fahy and William Furlong on the 12th of August, 1865, at seven p.m., when I heard a clap, like the shot out of a cannon, very quick

> and not like thunder; this was followed by a buzzing noise, which continued for about a quarter of an hour, when it came over our heads; and on looking up, we saw an object falling down in a slanting direction. We were frightened at its speed, which was so great that we could scarcely notice it; but after it fell, we proceeded to look for it, and found it at a distance of forty yards, half buried in the ground, where it had struck the top of a potato drill. We were some time in looking for it (a longer time than that during which we had heard the noise). On taking up the stone, we found it warm, milk warm, but not hot enough to be inconvenient. The next day it was given up to Lord Hawarden.

The total recovered mass from this fall was 2,270g. The meteorite has been classified as a H5 chondrite and is stored in the Natural History Museum, London.

1902 – Crumlin, County Antrim (54°36' N, 6°12' W)

The second meteorite to strike Ulster fell at 10.30 a.m. on 13 September 1902 in Crumlin, County Antrim. The location, some 32km west of Belfast city centre, is almost on the shores of Lough Neagh. It is quite probable that some remains of the Crumlin meteorite currently reside beneath the placid ripples of this large lough.

Below is an extract from W.E. Besley's account in the 1903 *Memoirs of the British Astronomical Association*:

> Three men were loading hay in a field at Crosshill, near Crumlin, County Antrim, when a noise like thunder or the rolling of drums broke overhead. One of them thought it was at Crumlin Mill, rather more than half a mile away, and described the report as twofold and followed by a whizzing noise or the sound of escaping steam. A second believed the cause was the running of a train off the line near by, and ran to look over the hedge, about a dozen yards off, returned, and put up a forkful of hay during the time the sound lasted. The former, from his position on the top of the haystack, saw something like a 'whirl' going into the ground about 70 yards off in the adjoining field (sown with

corn) with lightning speed. There was an explosion and the soil was thrown a considerable distance above the standing corn. When dug out the object, which had embedded itself in a straightdownward course for 13 inches, was found to be quite hot, continuing so for about an hour.

This fall even made it into *The Times* (11 November 1902):

On Saturday, September 13, at 10:30 a.m., a loud noise as of an explosion was heard at several places on the western side of Belfast – among others, Antrim, Crumlin, Lisburn, Moira, Lurgan and Pointzpass. It is noteworthy that of these places Antrim and Pointzpass are as far as 30 miles apart. An observer at Crosshill at first thought that the detonation was due to the bursting of the boiler in the mill at Crumlin, a mile away, while other persons state they heard a rattling noise, similar to that made by a reaping machine, but much louder. The detonation was followed by another sound, like that made by escaping steam.

A man, W.J. Adams, who was gathering apples from a tree on the edge of a cornfield at Crosshill, near Crumlin, was startled by these noises and still more so, immediately afterwards, by the sound as of a heavy object striking the ground not far from where he was standing. Seeing a cloud of dust rise above the growing corn, at a distance of about 20 yards from him, Adams rushed towards it, and found that a deep hole had been made in the soil. On reaching the spot he noticed a sulphurous odour around, and as to this he is corroborated by two other men, employed on the same farm, who were only a few yards further away.

Adams went for a spade, and, within a quarter of an hour of the fall, extracted from the hole a black dense stone, which had penetrated to a depth of a foot and a half and had then been stopped by a much larger stone lying in the soil. The black stone was hot when extracted, and is said to have been warm even an hour later.

The Crumlin meteorite also caused a political fireball when it was controversially, and surreptitiously, whisked away to London, resulting in a local and national outcry, complete with the publication of a bitingly satirical cartoon (in the 16 November 1902 edition of the *Irish Weekly Independence and Nation*) in which 'Pat', an Irish policeman, is sketched running after a fat 'John Bull' English collector of meteorites who is heading straight for the signposted British Museum clutching his spoils. This was the first, but not the last, controversy over an Irish meteorite.

The total recovered mass from this meteorite is 4,255g and, geologically, it is classified as an ordinary chondrite (more specifically, but not definitively, as an L5 chondrite for a specimen that was sold on eBay).

1969 – Bovedy, County Derry/Londonderry (54° 34'N, 6° 20'W)

At 9.22 p.m. on 25 April 1969, a fiery-white fireball, brighter than the full moon, scorched its way through the atmosphere above County Derry. This one actually caused damage to property: a smaller fragment of the fireball (some 513g) broke off and smashed through the asbestos roof of the Royal Ulster Constabulary Central Stores in Sprucefield, County Antrim, and then broke in two. Was this a Republican meteorite? The main mass (some 4,950g), however, inexorably continued on for a further 60km until it finally ploughed into Mr Samuel Gilmore's farm in Bovedy.

This fall is rare for several reasons, not least the fact that it is one of the very few meteors to be captured on audio. A woman who was outdoors that night with a microphone, trying to record nighttime birdsong, managed to get something more than hooting owls. As was summarised in a much later review of the incident in the *Londonderry Sentinel* on 16 February 2014, 'she caught the detonations through a microphone. The booms and rumblings echo around for several seconds after the first detonation, and a dog starts to bark soon after.'

The Bovedy meteorite is very special for six reasons. (1) It hit a building; (2) it was witnessed; (3) there are very few fragments, so specimens are scarce; (4) it was recorded on audio; (5) the main fragment is large by Irish and British standards; (6) the rock itself is of a relatively rare type.

Bovedy specimens are much sought after – very few are in private hands. The farmer, Samuel Gilmore, has passed away, but his son, Johnston, is very much alive. He was ten years old when the stone fell on his father's farm. And, as also noted in the *Londonderry Sentinel* article, 'Mr [Samuel] Gilmore's son admits that a piece of the meteorite remains in the ownership of the family, but that it will not be parted with, and added: "It is a good story, a good bit of family history."'

The total mass collected from this fall was 5,460g and it was classified as an L3 chondrite.

1999 – Leighlinbridge, County Carlow (52° 40'N, 6° 58'W)

The most recent of our intimate alien encounters took place at 10.10 p.m. on 28 November 1999, when a fireball shot over County Carlow and detonated in the atmosphere several times, leaving scattered fragments to be found at the side of the Leighlinstown to Bagnelstown road.

The Leighlinbridge meteorite had all the elements of high drama. The *Carlow People* newspaper reported that there was

> a brilliant exploding fireball … observed over Carlow town, which lasted several seconds before fading. Shortly afterwards, when the slower travelling sound waves reached the stunned observers, a loud detonation was heard and was described as 'massive' … it shook houses … some feared an explosion … people were too scared to go outside.

The Leighlinbridge meteorite seems to have exploded high in the atmosphere, and it is quite likely that other fragments are waiting to be discovered. But it won't be easy. This area of Ireland is not good for meteorite hunting. On the plus side, if you do find one, it can be sold for about €300 per gram.

Four stones were recovered in searches that went on until January 2000 (at least), one of the first being on 12 December 1999, by a local woman who was out walking her dog.

The Leighlinbridge meteorite also has the sulphurous whiff of controversy associated with it, almost reminiscent of that surrounding the Crumlin meteorite of 1902 (which was controversially appropriated by the Natural History Museum, London). The Leighlinbridge controversy involves Scottish-based meteorite collector Rob Elliott, who generously offered a personal reward of £20,000 to anyone finding a good-sized piece. They did, and (apparently) the fragment ended up in the British Natural History Museum, and not in an Irish museum. Cue row.

The total mass collected for this fall was 271g (total for all four stones) and, geologically, the meteorite is classified as an L6 chondrite.

And ... Who Died?

At the very start of this chapter I said that one of our aliens even killed a man. I could find no 'official' reports, but I did find the following curious excerpt from the National Library of Ireland catalogues concerning, of all things, a book on the rapemills of the Barony of Eglish, Birr, County Offaly:

> The ruins of this old rapemills is found on the Birr–Banagher Road about half way between these towns. How long the mill was in operation is uncertain but when Cooke did his agricultural survey of Offaly in 1801 he refers to rape as being one of the crops being grown in the Barony of Eglish. Cooke, in his History of Birr, tells of a remarkable event which took place at the mill in the 19th century. He says 'At that time a fireball or meteorite stone ... descended upon the mill and having broken through the roof, blew out all the windows, threw down the lofts, and killed Mr Woods, the Proprietor, who happened to be there at the time.' Mr Cooke also says 'the neighbourhood was remarkable for the appearance there in 1830 of white magpies – birds very uncommon in Ireland'. The letter-box on the wall of the old ruins bears the Queen Victoria crest.

A very odd report. Can anyone research this further?

Final Words – Fireball or Furball?

In recorded history, only 24 meteorites have ever been found in Ireland and Britain, and Ireland has had the proud honour of being hit by eight of them. And don't think that all our specimens are in the British Natural History Museum. Our own museums and universities hold many of them, too. An individual stone is often cut or divided into smaller pieces and these can be either sold privately or given to other institutions. As a result, pieces of Irish meteorites are now scattered worldwide.

Sadly, Ireland is not a country that is conducive to meteorite hunting. Apart from the difficulty of accessing certain tracts of farmland, the nature of the land itself could hardly be any worse: soft grasslands, bogs, woods, and thousands of rivers and lakes – all guaranteed to swallow any (small) meteorites without a trace.

Every year meteorite finds are reported by members of the public, and Nigel Monaghan of the National Museum of Ireland – Natural History is one of the people who is charged with verifying these reports. Unfortunately, however, only extremely rarely do any of these 'meteorites' turn out to be the genuine article. More often than not they are mere lumps of metal or a slightly odd-looking, but otherwise normal, rock. Occasionally, fossils get mistaken for meteorites, and, sometimes, the bizarre turns up. Nigel once had to tell someone that their treasured meteorite was, in fact, a cow's furball.

But hey ... if you don't look you certainly won't find. It's like the Lotto: you *could* be that little old lady in Leighlinbridge who bags twenty grand for picking up a golfball-sized lump of black crusty rock. Happy hunting!

9

In Awe of Aurum: Is Our Ore Not Our Ór?

Introduction

Gold. Actually finding some – even a minuscule amount – can fill a person full of joy and produce an involuntary adrenaline surge. I myself experienced such a rush when I found a small fleck of gold in the River Dodder at Bohernabreena (south Dublin) while on a gold-panning trip organised by the Irish Geological Association on a chilly but bright December day in 2012 (see Plate Figure 6). To the participants' initial bemusement, this trip was also being filmed by Ireland's national broadcaster, RTÉ. Apparently, RTÉ was looking for a much-needed happy item for the *Six One* television news to counteract the preponderance of gloomy economic stories that were then filling the country's living rooms. Not only did I find a fleck of gold – the very first in my entire life – but my reaction and accompanying babble (there was no rehearsal) was captured on camera and broadcast to the nation. And I have to say, it was quite something to go from being a hard-bitten geologist in his mid-forties to feeling like a kid of seven again. But that is what gold can do to you.

Irish gold. The public astutely ask two (related) questions. First: if we have gold then where is it? Second: where did the gold that was used to make our world-famous Early Bronze Age jewellery and ceremonial artefacts actually come from? I will attempt to answer those questions, using the most up-to-date research on the subject.

IS THERE NATIVE GOLD IN IRELAND?

Yes. The Irish have known about gold in Ireland for millennia. Getting at it has been the problem.

The Wicklow Gold Rush

Most historical gold that has been found has come from placer deposits: in other words, gold occurring as tiny flecks and up to fist-size nuggets found in streams and rivers. It was this type of gold that produced our only full-blown gold rush. This happened in 1795 in Wicklow, in the streams flowing through Ireland's 'Golden Triangle' – the area between Aughrim, Avoca and Croghan Kinshelagh. The gold rush was sparked off on 15 September by a 'common labourer' who was sweatily felling timber on the estate of Lord Carysfort, on the northern slopes of Croghan Kinshelagh. When the workman paused to inspect the soil and roots of one felled tree, something caught his eye: a shining half-ounce nugget of pure gold. It is safe to assume that he immediately abandoned his felling duties for Lord Carysfort and set about looking for further nuggets. Word then quickly spread among the poorer classes who, over the next number of days, began to converge on the area. By 11 October, there were some 4,000 people in the streams and embankments in and around Croghan Kinshelagh looking for gold.

Inevitably, all this activity drew the attention of the British authorities in Dublin Castle. The poor, albeit not many of them, were getting too wealthy, and the State was losing out. On 16 October, soldiers were dispatched from Dublin to march on Croghan and to claim the gold on behalf of His Majesty King George III. The peasants were to be sent back to their homes from where they could get on with their normal day jobs of subsistence survival and working for their rich landlords. Thus ended, after little more than a month, the popular wing of the Wicklow Gold Rush. After the peasants had been effectively disbanded, the area was then more professionally mined by the authorities – with a small interruption to squash the 1798 Rebellion. As Peadar McArdle wonderfully put it in his book *Gold Frenzy*, a garrison of troops remained in the area until roughly 1807, just in case any of the locals found themselves 'a bit distracted' again.

The Legendary Nugget

What of the legendary monster nugget that was found as part of the Wicklow Gold Rush? How big was it, and where is it today? Not easy questions to answer, particularly the second one. There is good evidence that it was a 22oz (624g) nugget of almost pure gold, that it was found in Ballin Stream, and that it was/is the largest gold nugget ever found in Europe. What happened to it? The best evidence suggests that it was purchased for £80 by two gentlemen named Abraham Coates and Turner Camac and they in turn presented it, in early 1796, to King George III. The king, in his infinite wisdom, then allegedly had it made into a snuffbox.

King George III is the royal whose epithet is 'Mad'. However, his troubled condition was most likely porphyria, not 'mental health' as such. And his health was certainly not helped by his physicians inadvertently poisoning him with medicaments of mercury, so one has to feel a bit sorry for him. At any rate, the original nugget seems to have been made into some sort of royal object and almost certainly no longer exists. Fortunately, several gold-painted lead models of the great nugget were made and these still do exist (see Plate Figure 6). We can but stare in awe at these and dream of what it would have been like to hold the real thing – or to have actually found it. Alternatively, we can admire the genuine, but much smaller, 3oz Wicklow gold nugget held by the National Museum of Ireland in Dublin.

The Source of the Wicklow Gold

A major mystery surrounding the Wicklow Gold Rush is the original source of the gold. True, it was panned and sieved from local streams, but this was not where the gold actually originated. Many professional geologists, and a goodly number of amateurs, have gone in search of the fabled mother lode – considered to be one or more bedrock quartz veins stuffed with gold just waiting to be hacked out. Sadly for the romantics, this idea is (almost) dead in the water. The latest geological evidence strongly suggests that the original gold occurs only as micron-sized particles within several of the rock types in the Avoca area: in the chalcopyrite of the copper deposits of Avoca and Kilmacoo; in the ironstones of the

Ballycoog–Moneyteige ridge; and as microscopic grains in quartz veins, utterly invisible to the naked eye.

These tiny bits of gold then weather out of their bedrock homes into the local streams where they aggregate around some nucleating particle, such as another bit of gold or a tiny grain of quartz, and grow over time to form flecks. If conditions are really favourable and there is physical amalgamation of these flecks, nuggets are formed.

Gold in the Rest of Ireland

There is gold in three of our four provinces. The following short list represents only the main occurrences; there are others. In Leinster, we have gold at the Goldmine River and at Kilmacoo, both in County Wicklow. In Connacht, there is gold at Croagh Patrick and at Cregganbaun, both in County Mayo. And in Ulster, there is gold at Cavanacaw and Curraghinalt, both in County Tyrone, and at Clontibret in County Monaghan.

The gold at Croagh Patrick is possibly the most remarkable. It is one of the very few localities in Ireland where gold can be seen with the naked eye within its original bedrock host (veined quartzites). A large boulder from the side of the mountain – with gold so obvious that anyone could see it – is on public display at the Geological Survey of Ireland offices at Beggar's Bush in Dublin. However, because of the religious and cultural significance of Croagh Patrick itself, no gold mining has been allowed. Which means that the gold is still there. If you are doing a barefoot pilgrimage, you might like to cast your eyes sideways on occasion and see if you get a golden glint reflected back at you. It may make the walk a little less painful.

The gold deposit at Cavanacaw is also remarkable in that it is the only operational gold mine in Ireland and has been producing Irish gold since 2007. The gold in this case is hosted in quartz veins and mostly as micro-grains within crystals of iron pyrite, a mineral better known as 'fool's gold'. However, in this particular case, there is a very fine line between the fool and the entrepreneur. It is gold from Cavanacaw that goes into a range of certified 18-carat Irish gold jewellery … and very popular it is, too.

Thus, there is not a shadow of a doubt that Ireland once had, and still very much does have, gold deposits. Gold has been found and exploited in the past, gold is being produced at the present, and there are known reserves of gold, if we choose to mine them, for the future.

So, you may ask, where did we get the gold used in our ancient Celtic artefacts?

Early Bronze Age Irish Gold Artefacts – Are They?

We have some superb ancient Irish gold jewellery. There are days when there are queues at the National Museum of Ireland – Archaeology to get in to see the fabulous gold pieces that are on display. The workmanship is breathtaking; the gold is dazzling. The fact that our ancient ancestors made these gold pieces can fill us with a confidently understated sense of national pride.

Our gold pieces date from two archaeological periods. First, the Chalcolithic Period (*c*.2500 BC to 2100 BC), from which we have found rare gold discs, five gold bands from County Cavan, and three basket ornaments, one of which is probably the earliest gold artefact in Ireland (although it seems to have actually originated in Iberia). Second, the Early Bronze Age (*c*.2100 BC to 1550 BC), a period that has given us, among other items, some 87 lunulae (probable neck ornaments decorated with geometric designs). The lunulae are usually assumed to have been made using Irish gold and, based on their chemical composition, a source for the gold around the Mourne Mountains has been proposed, but not confidently demonstrated.

Thus, a question surrounds the origin of our Early Bronze Age gold artefacts, which must also include all the treasures that were looted and pillaged over the ages and that no longer exist. Just where did all the gold come from?

To clear up one aspect: all the experts do seem to agree that almost all of the lunulae and so on were 'made in Ireland' by the 'native Irish' and were not imported as ready-made objects. They seem indigenous. The question, therefore, reduces to: Where in Ireland did they get the gold from?

Chris Standish, formerly of Bristol University, has done some very interesting work to help answer that question. As I write, his results are hot off the press and will unquestionably be subjected to intense scrutiny. He managed to get permission to do an innovative type of analysis on our ancient gold artefacts: not a traditional elemental analysis, but looking at isotopes of lead. Almost all gold contains some lead and that lead comes in several different isotopic varieties; different 'flavours' of lead, if you will. And Chris had the means to detect the flavours of lead in a gold artefact: to determine its so-called 'lead isotopic signature.' If we know what 'flavour' is in the gold, and we know what 'flavours' are in various potential source areas, then we can – to keep the metaphor going – lick the gold, lick the potential source areas, and see if they taste the same. This is essentially what Chris did. By comparing the lead isotope characteristics of the gold artefacts with the lead isotope characteristics of bedrock and placer gold in a selection of Irish, British and Continental European gold deposits, he was able to narrow down the source of the gold used in the earliest Bronze Age Irish gold ornaments.

He has made the following qualified conclusions. The gold did not come from the west of Ireland (e.g. the known Mayo occurrences today), nor from the northern part of Ireland (the Tyrone and Monaghan occurrences). It did not come from the Mourne Mountains, which is where many archaeologists think it came from. It did not come from southeast Ireland, which would include the Wicklow placer gold deposits of the later gold rush. The southeast had previously been ruled out on the basis of elemental analysis, anyway. Neither did it come from the southwest of Ireland, where gold is almost non-existent. The 'flavour' of the gold artefacts, when both the new isotopic data and pre-existing elemental data are combined, is not a match for the 'flavours' of any of these localities or provinces. And by now, as you might have spotted, we have run out of Ireland. Which means ... that the gold itself probably did not come from Ireland. To go back to our analogy, the Irish gold artefacts do not 'taste' of Ireland.

Chris Standish is actively working to pin down with confidence where the Early Bronze Age gold actually did come from. He suggests that there is one leading contender for a source; one general area to where all the currently available evidence – chemistry, isotope data and circumstantial historical/archaeological data – points its provisional finger. Cornwall. In southern Britain.

If further research backs this up, it opens a whole new set of questions about just who was mining the gold, who was using it, how it was getting from supplier to user, and, not least, how come the native Irish were not using their own native Irish gold?

10

The Irish Mineral Connection: Introduction, Prelude and Fugue

Introduction

I first fell in love with minerals when I was seven years old. I had just found a small, but perfectly formed, water-clear crystal of quartz in Galway ... and I thought I had found a diamond. My delight remained unbounded even when I learned a week later that it was not a diamond. That little quartz discovery was responsible for kick-starting my lifelong love of geology.

To many, a 'mineral' (or 'miner'tle' as it's occasionally pronounced in certain localities) means, first and foremost, a fizzy drink. To me – even now – whenever I see an advertisement for 'minerals' I automatically think of crystals. Endless disappointment while doing the grocery shopping.

This chapter is not a mineralogy of Ireland, but something a bit different. It is a bird's-eye view of minerals that have an Irish connection, i.e. all minerals, found either in Ireland or elsewhere in the world, that have been named after an Irish person or an Irish place or that were first described here. I have included currently valid mineral species and also, for historical purposes, now-discredited species.

In writing this chapter I want to express my immense gratitude to Andrew Tindle of the Open University (UK) for publishing his monumental tome *Minerals of Britain and Ireland*. This chapter is based on material that he collated, although I have made various modifications, corrections, additions and side comments.

Prelude

A few terms need to be introduced in order to better appreciate the significance of some of the minerals entries below.

- *Type locality*: This is the location where a mineral was first discovered. It is the reference 'ground zero' for all other occurrences of that mineral species, wherever else in the world it might turn up.
- *Discredited mineral*: This is a mineral that had been given a name, but was subsequently found not to be a proper mineral under the modern rules for defining or naming a mineral. Reasons for this include that it is a mixture of minerals (so not a single mineral species), it does not have a well-defined chemical composition, or it had been described earlier under a different name (in which case the earlier name takes precedence). But, once discredited, a mineral name can never be officially used again. And I agree with that. Nevertheless, I do find it sad that there can now never be a 'killinite,' as named after Killiney (south County Dublin), or a 'haughtonite,' named after Trinity College Dublin geology professor Samuel Haughton. These once-common, nineteenth-century mineral names are now in the bin of history. On the brighter side, if one finds a new mineral and wants to honour these places or people anew, one can still do it: in principle, there's nothing to stop you calling your new mineral 'killineyite' or 'samhaughtonite.'
- *Naming of minerals*: In essence, a mineral, whether discovered and named a millennium ago or yesterday, is now internationally considered to be valid if it has passed muster with those strict nomenclaturial people at the deliciously non-succinctly named International Mineralogical Association – Commission on New Minerals and Mineral Names (IMA–CNMMN), later changed to the Commission on New Minerals, Nomenclature and Classification (the partly palindromic abbreviation IMA-CNMNC). The rules and regulations are intricate. But the most important rule is that any mineral must have a well-defined

(though possibly between certain limits) structure and chemical composition.

FUGUE

Mineral Species (Valid) Named after Irish Locations

Corkite: $PbFe^{3+}_3(SO_4)(PO_4)(OH)_6$. The only mineral to be named after an Irish county: County Cork, with a type locality at Glandore iron mine. At Glandore itself, corkite forms small brown translucent crystals with a brilliant lustre and occurs associated with quartz in a gossan matrix (the top oxidised part of a mineral deposit). Corkite also occurs in cavities in massive goethite as aggregates of minute yellow rhombs. In addition, Ryback and Francis (1991) report it from nearby Aghatubrid Beg mine.

Garronite: $NaCa_{2.5}(A_{16}Si_{10})_{32}.14H_2O$. Named after the Garron Plateau in County Antrim, with a type locality at the western side of Glenariff Valley. This mineral is a type of zeolite (a type of silicate with a 3D structure and lots of water). Garronite is rare worldwide but is locally common in some of the amygdales ('almond-shaped holes') within the Antrim basalts. It forms radiating white to colourless aggregates in association with other zeolites. In Ireland, garronite has been reported from counties Antrim and Derry.

Gobbinsite: $Na_5(Si_{11}A_{15})O_{32}.11H_2O$. Named after the Gobbins, Island Magee, County Antrim, and with a type locality at the coastal escarpment near Hills Port, south of the Gobbins. An extremely rare mineral of the zeolite group, it occurs as chalky white to pale-pink fibrous masses within Antrim basalt amygdales, notably at the Gobbins and at Magheramourne quarry, near Larne, County Antrim.

Gortdrumite: $(Cu,Fe)_6Hg_2S_5$. Named after Gortdrum mine in County Tipperary. Gortdrumite is a rare, mercury-bearing mineral that was found in a vein that contained dolomite and barite and that cut across the local dolomitic limestone in Gortdrum. The mineral occurs as blackish-grey grains (prisms), is relatively soft, and was only discovered in 1983. It probably crystallised at quite low temperatures (100–200 degrees Celsius).

Killalaite: $Ca_3Si_2O_7.H_2O$. Named after Killala Bay, near Inishcrone, County Sligo, and with a type locality there. Killala Bay is well known to Irish geologists as having a very large gabbro dyke exposed on its western shore and for being the best fjord in the country. But there are also basalt/dolerite dykes in the area. The mineral killalaite was formed from the heat of these latter dykes altering local limestone, with killalaite forming in cavities and veins therein. Crystals are colourless, less than 2mm long, and have a characteristic bow-tie shape to them. A second occurrence of killalaite has been reported from Carneal, Larne, County Antrim.

Larnite: Ca_2SiO_4. Named after Larne in County Antrim, and with a type locality at Scawt Hill, near Larne. The geology of Antrim is fundamentally unusual: in essence, it is a Cretaceous chalk base smothered in a thick topping of Paleogene basalt. The basalt, a silica-rich, iron-bearing hot magma, came up (mostly) through dykes, and these hot dykes often cut through the underlying chalk. When they did, the chalk, a calcium carbonate rock, was metamorphosed and altered in various ways by the basalt, a silica-rich iron-bearing hot magma. The result of this meeting of cool calcium with hot silica is a range of minerals that contain these elements as essential ingredients, like larnite. Larnite itself is not easy to see with the naked eye, forming fine-grained granular masses. It can also form by a very different process, as an alteration rind on some nodules of flint.

Scawtite: $Ca_7Si_6O_{18}(CO_3).2H_2O$. Named after Scawt Hill, Larne, County Antrim, and with a type locality there. Scawtite is formed in essentially the same way as larnite (above), only in this case it is formed in chalk blocks that had been broken off by dolerite (basalt) dykes and had then become trapped and metamorphosed within the hot, surging magma. The locality of Scawt Hill is famous for this, and a number of new minerals have been identified from this type of occurrence. Scawtite occurs as 1.5mm colourless platy crystals or as tiny (0.03mm) fibrous aggregates of radiating flakes. Scawtite has also been found at Ballycraigy (Larne) and at Carneal (Larne) from a similar style of formation.

Mineral Species (Valid) with Irish Type Localities, but without Irish Citizen or Locality Names

Bredigite: $Ca_7Mg(SiO_4)_4$. The type locality is at Scawt Hill, Larne, County Antrim. This is a rare mineral, but one that formed in fundamentally the same way as scawtite and larnite: interaction between hot dolerite and cool chalk. Bredigite forms dark reddish-brown crystals, less than 1mm across, that can look (pseudo) hexagonal one way and barrel/boat-shaped another. According to the authors who described it first (Tilley and Vincent, 1948), the type specimen itself was discovered in a slag.

Gmelinite: $(Ca_{0.5},Na,K)_4(Si_8Al_4)O_{24}.11H_2O$. Not a single mineral but, better still, a whole mineral series within the zeolite group. The type locality is at Little Deer Park, Glenarm, County Antrim (with a co-type in Vicenza, Italy). This series includes gmelinite-Ca and gmelinite-Na, both of which are found in a quarry on the south side of Whitehead, near Island Magee, County Antrim. Gmelinite is not common, but can be found in a narrow zone along the eastern coast of Antrim. Gmelinite is a gorgeous, translucent orange-pink hexagonal mineral, occurring in clusters at least 1.5cm long, that look like mouth-watering, six-sided orange-sorbet ice creams. Go find some!

Hydrocalumite: $Ca_4Al_2(OH)_{12}(Cl,CO_3,OH)_{2-x}.4H_2O$. The type locality for hydrocalumite is Scawt Hill, Larne, County Antrim. This is a rare mineral that is found in vugs in larnite-bearing rock formed as a result of the interaction between chalk and dolerite (basalt cooled in a dyke) at the classic locality of Scawt Hill. The mineral occurs as colourless to light-green crystals that have a very good (perfect, in fact) cleavage. The name, in this case, is a simple allusion to its fundamental chemistry: water (**hydro**), **cal**cium and **alumin**ium.

Osumilite-(Mg): $KMg_2(Al,Fe)_3(Si,Al)_{12}O_{30}$. The type locality is at Tieveragh, near Cushendall, County Antrim. Tieveragh is a classic locality in the mould of Scawt Hill, only in this case the obligatory dolerite dyke intrudes and alters a sandstone, rather than a chalk. Osumilite-(Mg) only occurs as microscopic (100 micrometre) crystals in the fused sandstone, so you won't see it with the naked eye. The name itself has not yet been fully approved by the IMA-CNMNC.

Portlandite: $Ca(OH)_2$. The type locality is at Scawt Hill, Larne, County Antrim. It is another rare mineral that was identified from the dolerite/chalk interaction at Scawt Hill. Portlandite occurs as colourless hexagonal platy crystals that can be found lining small vugs. The name is derived from the fact that Portland cement, upon hydration, produces this compound. The compound, when it happens in cement itself, is not called a mineral, but a chemical, because the compound derives from man-made materials: minerals, *sensu stricto*, must be produced by Mother Nature.

Rankinite: $Ca_3Si_2O_7$. The type locality is at Scawt Hill, Larne, County Antrim. It is another of the rare calcium/silicon minerals that was identified from the dolerite/chalk interaction at Scawt Hill. Rankinite forms small colourless to dark-grey masses, with no obvious defined shape or cleavage. Rankinite has also been found as part of the basalt/limestone interaction assemblage of minerals at Killala Bay, near Inishcrone, County Sligo.

Vaterite: $CaCO_3$. The type locality is at Ballycraigy, Larne, County Antrim. Vaterite is the rare hexagonal form of otherwise common calcium carbonate. It is chemically identical but structurally different to the calcium carbonate mineral aragonite, which is orthorhombic, and calcite, which is trigonal. This mineral has an unusual occurrence: it has formed within a carbonated calcium silicate hydrogel and appears as very finely fibrous or platy crystals. But if you want to collect some, you must be aware of one thing: it is unstable in water at room temperature – this will cause it to dissolve and reform as normal calcite, which presents some issues when preserving it.

Mineral Species (Valid) Named after Irish Citizens (or Those with a Strong Connection to Ireland) but That Are Not Found in Ireland

Apjohnite: $Mn^{2+}Al_2(SO_4)_4.22H_2O$. Named after the Trinity College Dublin chemist and mineralogist, James Apjohn (1796–1886), who was born in Grean, County Limerick. The type locality for apjohnite, however, is in Mozambique. The mineral is very soft, occurs as needle-like white to clear crystals, rarely more than 5mm in length, and is soluble in water. Rather oddly, as it turns out, apjohnite has been recorded in Ireland: in the supplement to Greg and Lettsom (1858). But this latter report cannot be apjohnite because it describes a brown metallic ore mixed with pyrite. Thus, at the moment, no apjohnite has been found in Ireland.

Bernalite: $Fe(OH)_3.nH_2O$ (n = 0.0–0.25). Named after the lecturer in structural crystallography and assistant director of research at the University of Cambridge, and later professor of physics at Birbeck College, University of London, John Desmond Bernal (1901–1971). Despite often being cited as British, Bernal was born in Brookwatson, Nenagh, County Tipperary. He had an exotic mix of ancestry: Irish, Italian, and Iberian Sephardic Jewish. Bernalite is a very unusual bottle-green iron hydroxide with an adamantine to resinous lustre, and was originally found at Broken Hill, New South Wales, Australia.

Chenevixite: $Cu^{+2}{}_2Fe^{+3}{}_2(AsO_4)_2(OH)_4.(H_2O)$. This is an uncommon arsenate mineral, so far only fully confirmed from Cornwall (in the context of Britain and Ireland; the type locality is at Wheal Gorland, near Redruth) and is named after Richard Chenevix (1774–1830). Chenevix was an Irish chemist and mineralogist of Huguenot descent (he was fourth-generation Irish). His great-great-grandfather came to Ireland from Limay (France) and settled in Portarlington, County Laois. Richard Chenevix had a distinguished academic career and led a very colourful life: he was married but had two illegitimate children and he was a veritable pit bull in scientific 'discourse'. However, contrary to many authoritative works (the *Oxford Dictionary of National Biography* among them), he was not born in or near Dublin: he was born in Ballycommon, County Offaly.

Crawfordite: $Na_3Sr(CO_3)(PO_4)$. Named after the co-discoverer of the element strontium and Edinburgh-based physician and chemist, Sir Adair Crawford (1748–1795). Crawford was born in Crumlin, County Antrim. Crawfordite was named in 1994 for a soft, colourless mineral that was first found in the Kola Peninsula of Russia.

Emeleusite: $Li_2Na_4Fe_2Si_{12}O_{30}$. Named after Belfast native Charles Henry Emeleus (born 1930), who went on to have a great career, including being made Reader in the Department of Earth Sciences, University of Durham. Emeleusite is a very rare pale-brown mineral from Igdlutalik, Ilimaussaq, Greenland.

Lonsdaleite: C. That 'C' is for pure carbon. Lonsdaleite is a rare allotrope of carbon, found almost exclusively in meteorites, that is based on a hexagonal crystal pattern (graphite and diamond are some other allotropes of carbon based on layered and cubic patterns, respectively). Named after the eminent Irish crystallographer and professor of chemistry at the University of London, Dame Kathleen Lonsdale (née Yardley) (1903–1971). Kathleen Yardley was born in Newbridge, County Kildare, and is a fantastic example of an Irishwoman reaching the pinnacle of her profession. The type specimen of lonsdaleite comes from the Canyon Diablo meteorite, and although the original material contained impurities that affected its measured properties, artificially produced pure lonsdaleite turns out to be even harder than diamond.

Mackayite: $Fe^{3+}(Te^{4+}_2O_5)(OH)$. Named after the Dublin-born Irish-American mine operator on the well-known gold and silver deposit of the Comstock Lode (Nevada), John William Mackay (1831–1902). Mackayite is a lovely olive-green mineral that occurs as four-sided pyramids and was first described from the Mohawk Mine, Nevada, United States.

Manganbabingtonite and **Scandiobabingtonite**: $Ca_2(Mn,Fe)FeSi_5O_{14}(OH)$ and $Ca_2(Fe^{2+},Mn)ScSi_5O_{14}(OH)$. Named after the physicist, mineralogist and founding member of the Geological Society of London, William Babington (1756–1833). Babington was born in Port Glenone, near Coleraine, County Antrim. He first curated and then bought the huge mineral collection of the Third Earl of Bute, John Stuart. Manganbabingtonite is a dark-green, translucent prismatic mineral and was first described from the Rudnyi Kaskad mineral deposit in Eastern Siberia, Russia (excellent specimens later turning up in China). Scandiobabingtonite forms moderately hard, colourless to dark-green small crystals, occurs as an accessory mineral in granite, and was first found at the Seula Mine in Piedmont, Italy.

Oldhamite: (Ca,Mg)S. Named after the Dublin-born Irish geologist and later director of the Indian Geological Survey, Thomas Oldham (1816–1878). Oldhamite is a rare mineral, but is a close structural cousin to the very common galena (PbS). It occurs in Poland and Germany as tiny dark-grey crystals (often as inclusions in other minerals), as well as in America and Asia. The type specimen actually came from the Bustee meteorite found in India in the nineteenth century.

Simpsonite: $Al_4(Ta,Nb)_3O_{13}(OH,F)$. Named in honour of Australian government mineralogist and analyst Edward Sydney Simpson (1875–1939). Simpson was born in Woollahra, Sydney, of English and Irish parents – the Simpson Desert is also named after him. Simpsonite is a hard, yellow to light-reddish brown mineral first described from the Tabba Tabba pegmatite in Western Australia.

Stokesite: $CaSnSi_3O_9.2H_2O$. This very rare hydrated calcium–tin silicate was named after the great Irish mathematician and physicist, Sir George Gabriel Stokes (1819–1903), who held the celebrated position of Lucasian Professor of Mathematics at Cambridge University (as did Sir Isaac Newton and as does Stephen Hawking) and who was a president of the Royal Society.

George Stokes was born in Skreen, County Sligo. The mineral itself has only been found in situ in Cornwall, the type locality being at Wheal Cock Zawn, Roscommon Cliffs, St Just, Cornwall. The Cornish crystals tend to be less than 7mm in length, are clear to pale-pink pyramids, and occur either as single crystals or as radiating groups.

Taaffeite (more correctly written, but much more difficult to read aloud, as magnesiotaaffeite-2N'2S): $Mg_3Al_8BeO_{16}$. Named after Count Edward Charles Richard Taaffe (1898–1967). The Count was a Dublin-based gemmologist who had been born in Bohemia (Czechoslovakia). His father, however, was the twelfth Viscount of Corran, Baron of Ballymote, County Sligo. The type locality for taafeite is Niriella village, Ratnapura, Sri Lanka. The history of the discovery of taafeite can be read in Chapter 11 of this book.

Mineral Species (Invalid) and Varieties Named after Irish (or British) Citizens

The technical nationality of a person in Ireland or Britain – or the United Kingdom of Great Britain and Ireland, as it became known when the Act of Union was passed in 1800 – is an interesting topic for debate, involving long and complex historical and genetic arguments. For the purpose of this section, I am just taking people born on the island of Ireland at any time or who worked here on a long-term basis. I leave to others any debate on whether they were Irish, English or British: I am taking the following minerals as Irish in the widest sense.

Andrewsite: Named after the medical doctor, fellow of the Royal Societies of Edinburgh and London, and professor at Queen's College Belfast, Thomas Andrews (1813–1885). Andrews was born in Belfast. The original andrewsite was discredited because it was found not to be a single mineral but a mixture of four relatively obscure, but true, minerals: barbosalite, chalcosiderite, hentschelite and rockbridgeite.

Cotterite: Named by a nineteenth-century professor of geology at Queen's College Cork (now UCC), Richard Harkness, in 1878 after the mysterious 'Miss Cotter' who found the original specimens in a quarry in her (very large) garden in Rockforest, near Mallow, County Cork. Cotterite is a variety of quartz where the crystals have a very unusual pearly metallic lustre to them. However, the name is a varietal name, not a proper full mineral name, making it analogous to 'amethyst' or 'citrine', for example. For more, see Chapter 16 and Plate Figure 7.

Doranite: Named after a nineteenth-century mineral dealer, Patrick Doran (1779–1880). Doran was born in Glassdrummond, County Down, and seems to have gone by the moniker 'Diamond Pat'. The type locality for doranite was Carrickfergus in County Antrim. But alas for Patrick Doran and his descendents, his mineral name is no more because doranite was found to be a magnesium-rich analcime, an already accounted-for member of the zeolite group.

Haughtonite: Named after a nineteenth-century professor of geology at Trinity College Dublin, the Reverend Samuel Haughton (1821–1897: also see Chapter 14). Haughton was born in County Carlow and was one of those great nineteenth-century polymaths who were truly excellent at many things: from geology to medicine, and from zoology to the science of humanely hanging people. Haughtonite is a mineral often mentioned in nineteenth-century papers on granite as being one of the dark micas. But it was discredited when later, more detailed, analyses revealed it to be just a normal variety of the dark and already named mica, biotite.

Hullite: Named after the stratigrapher, economic geologist, director of the Irish Geological Survey (Dublin) and professor at the Royal Dublin College of Science, Edward Hull (1829–1917). Hull was born in Antrim town, and the type specimen of hullite was from Carnmoney Hill, near Belfast, County Antrim. Although Hull's place in the history of Irish geology is quite secure, his mineralogical posterity took a hit when it was found that hullite was too vaguely described and could be either chlorophaeite or a magnesium-rich chamosite (a mineral of the chlorite group). Either way, 'hullite' was no more.

Kirwanite: Named after one of Ireland's greatest (and most eccentric) scientists, Richard Kirwan (1733–1812). Kirwan was born in Cloghballymore House, near Kinvara, County Galway, and was arguably Ireland's first mineralogist; he was also a chemist (mineralogy and chemistry went hand in hand in the old days), a prodigious inspector of His Majesty's Mines in Ireland, and a fellow of both the Royal Societies of London and of Edinburgh. The mineral kirwanite itself was poorly described and located ('north-east coast of Ireland'), and it has been said to be either a chlorite-like mineral, a synonym for glauconite, or some type of amphibole. The best guess by Andrew Tindle (2008) is that kirwanite is actually a ferrohornblende. With such konfusion, kirwanite is kaput.

Lehuntite: Named after the highly elusive personage of 'Captain Lehunt'. Who was he? Nobody knows. But not knowing the dedicatee doesn't discredit a mineral name. The type locality was at Glenarm, County Antrim, and the name gets thrown into the fires of oblivion because 'lehuntite' turned out to be a synonym of the zeolite mineral natrolite.

Loganite: Named after Sir William Edmond Logan (1798–1875) who has no Irish connection: he was born in Montreal, Canada, educated in Edinburgh (he had Scottish parents), and became the first director of the Geological Survey of Canada. Loganite was named in honour of Logan by Professor Richard Harkness (of Queen's College Cork, now University College Cork) in 1866 when he found what he thought was a new mineral at Lissoughter, County Galway. But loganite turned out to be a pseudomorph of pennine after amphibole (i.e. an amphibole mineral later geologically transformed to pennine while keeping the original amphibole shape); or possibly a mixture of the three minerals actinolite, diopside and talc. So ended loganite's run.

Scoulerite: Named after John Scouler (1804–1871), a surgeon, botanist, naturalist, professor of geology, natural history and mineralogy at the Andersonian University in Glasgow, and professor of mineralogy of the Royal Society, Dublin. The type locality was at Portrush, County Antrim. The name was frowned upon and discredited when it was realised that scoulerite was just a variety of the zeolite mineral thomsonite.

Mineral Species (Invalid) and Mineral Varieties with an Irish Type Locality, but Not a Citizen or Locality Name

Alumyte: The type locality is at Glenarm and Ballintoy, County Antrim. However, alumyte is but a synonym for metahalloysite, itself a synonym for halloysite-7Å, which is a dehydrated version of halloysite-10Å. So 'alumyte' isn't valid.

Chalilite: The type locality is in the Donegore Mountains, Sandybrae, County Antrim. But this former mineral is a substance of uncertain composition, possibly a variety of the zeolite thomsonite. Discredited.

Crucilite: The type locality is at Clonmel, County Waterford. Crucilite turned out to be a pseudomorph of hematite after arsenopyrite, i.e. original crystals of arsenopyrite were completely replaced later with the mineral hematite, but the hematite retained the shape of the arsenopyrite. Not an independent mineral, therefore discredited.

Harringtonite: The type locality is at Portrush and the Skerries, County Antrim. There is some suggestion that this mineral was named in honour of Robert Harrington (1751–1837) who was a surgeon and science writer from Carlisle, Cumbria. In any case, the tough white mineral became invalid when it was discovered to be just a mixture of two zeolites: thomsonite and mesolite.

Koodilite: The type locality is somewhere in County Down, and, like many invalid Irish minerals, turned out to be a variety of the zeolite mineral thomsonite.

Mesolitine: The type locality is at Newtown Crumlin, Clough, County Antrim. Yet another synonym of the zeolite thomsonite and yet another discredited species.

Parastilbite: The type locality is at Kilkeel, County Down. This is discredited by the IMA-CNMNC because it is really a synonym of either stilbite or epistilbite.

Plinthite: The type locality is somewhere in County Antrim, and the name is sometimes spelled 'plynthite'. This one is a real ratbag of a discredited mineral. In its time, the name has been applied to a mixture of analcime, hematite and montmorillonite; also to a mixture of the zeolites thomsonite and chabazite; and to a mixture of thomsonite and hematite. No question about it, this name had to go.

Rhodalite: The type locality is in County Antrim. Rhodalite, which is a really beautiful name and deserved better, was appended to an ill-defined clay mineral (whose definition even at the time was as clear as mud) and so had to be later discredited.

Silicite: The type locality is in County Antrim. The mineral was discredited because it was ordinary labradorite, a very well-known, calcium-rich member of the plagioclase feldspar group.

Talcite: The type locality is in County Wicklow. The name 'talcite' was doomed not only because it was found to be a synonym for damourite – which itself is not a valid mineral name either but a varietal name for a fine-grained, compact, greasy-feel version of muscovite – but it was also used as a synonym for the real mineral of talc itself. Slip the name into the bin.

Verrucite: The type locality is at Sandy Braes, County Antrim. Described by Andrew Tindle (2008) in his book as 'reddish-brown, compact and warty' (hence the name). Verrucite turned out to be a variety of the zeolite mineral mesolite. Bazooka that verrucite.

Mineral Species (Invalid) and Varieties Named after Irish Locations

Antrimolite: Named after County Antrim, and with a type locality at Ballintoy, west of Ballycastle. This mineral was discredited because it was found to be a variety either of thomsonite or of mesolite associated with calcite.

Erinite: Named after the island of Ireland: Erin. The type locality was at the Giant's Causeway, County Antrim. This mineral was badly defined originally and now seems to have been the mineral nontronite. However, the name 'erinite' has also been applied to chalcophyllite, cornwallite, monazite-(Ce), and an yttrium-rich pyrope garnet. Six synonyms! It had to go. A really great mineral name, though.

Kilbrickenite: Named after Kilbricken in County Clare, with a type locality at Monanoe (Kilbricken) mine, Quin, County Clare. It turned out that kilbrickenite was a local synonym for geocronite, and so was discredited.

Killinite: Named in 1818 after Killiney in south County Dublin, with a type locality there. Killinite had quite a prestigious history in the nineteenth century – it was widely mentioned in not only geology and mineralogy texts, but in the popular press also. Every natural history guide to Dublin seemed to mention it. And it got into several editions of the nineteenth-century mineralogical 'Bible', *Dana's Mineralogy*.

Dubliners in particular were very proud to have a mineral named after their local beauty spot. I remember as an undergraduate in Trinity in the mid-1980s that the name was still being mentioned. Alas, suspicions had been raised even by the mid-nineteenth century as to whether it was really a proper mineral. The nail in the coffin came in the 1980s when modern analyses of original specimens revealed them to be a variety of muscovite that had formed as a result of the alteration of a pre-existing mineral, spodumene. Very sad.

Kilmacooite: Named after Kilmacoo, County Wicklow, with a type locality at the Magpie and Connary Mines, Avoca, County Wicklow. This mineral is one of those that turned out to be an undefined mixture of two lead and zinc minerals. Name shafted.

Mornite: Named, allowing for an oddity in spelling, after the majestic Mourne Mountains, County Down, with a type locality there. Sadly, the name was swept down to the sea and drowned when somebody discovered that mornite was just a synonym for the well-known plagioclase feldspar labradorite. Down felt down; Mourne was in mourning.

Rosstrevorite: Named after Rosstrevor, County Down, with a type locality there. Rosstrevorite turned out to be a name given to a stellate (star-shaped) variety of epidote. And the crucial thing here is that it is *epidote*, an already named mineral; the good folks at the IMA-CNMNC do not allow full mineral name status just for different shapes of the one mineral.

Wicklowite: Named after County Wicklow, with a type locality there. This is an odd one. The name wicklowite seems to have been a bastardisation of the mineral name vichlovite, and this name itself does not seem to be a valid mineral name now. Wicklowite (vichlovite) is possibly some sort of lead vanadate, but its mineral status is described by Andrew Tindle (2008) as 'dubious', and the whole issue is wreathed in so much uncertainty that it is no surprise that the name was discredited. A great pity – I'd love there to be a valid 'wicklowite', but now it can never be.

Mineral Species (Invalid) Named after Irish Citizens (or Those with a Close Connection to Ireland), but Not Found in Britain or Ireland

Gieseckite: Named in honour of a former professor of mineralogy at Trinity College Dublin, Sir Charles Lewis Giesecke, formerly known as Johann Georg Metzler (1761–1833), who was born in Augsburg, Bavaria. Gieseckite was described as a greenish-grey mineral occurring as six-sided prisms and having a greasy lustre. Discredited because it is probably just a variety of the also discredited mineral elaeolite, which itself is thought to be a form of nepheline.

Kaneite: Named after the Dubliner, chemist, president of Queen's College Cork and author of *The Industrial Resources of Ireland* (1844), Sir Robert John Kane (1809–1890). This mineral was described from Saxony, Germany, and is said to have the chemical formula MnAs (manganese arsenide) but has had a doubtful history and is not confirmed. However, it is the only IMA-CNMNC non-approved mineral in this chapter that could, just possibly, yet be validated.

11

THE EMERALD ISLE? IRISH GEMSTONES

INTRODUCTION

Has Ireland got any emeralds? Has Ireland got any gemstones? This chapter did not turn out as expected.

Initially, I had intended to write about Irish gemstones: gemstones being defined as minerals that are partially or wholly faceted, polished and either left unmounted as collectors' pieces or mounted in gold, silver or platinum and sold as fine jewellery. This would be a reasonable definition of a gemstone, separating such high-grade material from more normal mineral specimens and eliminating things like tumbled minerals – rough pieces that have been polished in a mineral tumbler and maybe mounted in a simple way using silver wire – or natural crystals mounted as pendants. Tumbled and polished minerals can be produced by anyone and can be lovely. But they are not fine gemstones.

From the outset, I suspected that the topic of Irish fine gems and jewellery would be limited. I was genuinely shocked by what I found. I interviewed possibly the most eminent jewellery expert in Ireland – John D'Arcy, who, at the time of writing, owns Lapis Jewellers on Nassau Street in Dublin – and he in turn contacted many of the people he knows in the trade, including fellow jewellers, mineral collectors and gemcutters, both in Ireland and in England. I made additional enquiries myself among the geological community and other Irish jewellery-shop owners. But to both my and John's great surprise, neither of us could find one – *not even one* – bona fide example of a fine-cut gemstone that had originated from Ireland. None of the experienced gemcutters recalled ever having been given an Irish stone to facet. Well ... with just one exception. Some very fine mineral specimens of sphalerite from Silvermines (County Tipperary) were once submitted for faceting, but the gemcutters refused to cut them on the grounds that they were more valuable as natural specimens than as faceted stones. Extraordinary.

So, if you, dear reader, happen to know of an authentic Irish gemstone cut in modern times, please contact either myself or John D'Arcy. We would very much like to know of it.

Thus, having found that there are possibly no modern Irish gemstones, I had no choice but to broaden the scope of the chapter: Irish semi-precious minerals and organic 'gems'.

Ireland's (Semi-)Precious Minerals: A Selection Box

What follows is a confection of Irish minerals that are normally considered precious or semi-precious. I have tried, where possible, to follow the original theme of the chapter so that the mineral in question has been (or apparently has been) used in some sort of jewellery or other type of ornament.

Amber

Amber is not a true mineral: it is fossilised tree resin. For popular purposes, however, it is often treated as a mineral and a semi-precious organic gem. Amber has been used in Irish jewellery since at least the Bronze Age – see the amber necklace found with the Meenwaun hoard from County Offaly – and the assumption has been that the elite of the day controlled its import and use in Ireland. But not all the amber may be exotic to Ireland. There are reports of amber of a rich yellow colour being found on the coast near Howth, County Dublin, and amber has the potential to be found on any of the northern and eastern Irish coasts, although it won't be as common as on some of the southern or eastern coasts of England. The amber-like, ill-defined organic mineral called retinite has been found in coal seams on Rathlin Island, off the

Antrim coast. It is possible that some ancient Irish amber beads were made from amber discovered on (in) these shores. As I write, Lisa Moloney (Institute of Technology, Sligo) is trying to decipher the provenance of the Irish ambers.

Amethyst: SiO_2

Amethyst is the purple variety of quartz that contains some iron to give it the purple colour. It can be found in small amounts in several Irish localities, but two really stand out. The first is Achill Island, County Mayo – undoubtedly the most famous, and one of the oldest, of Ireland's quartz-collecting localities. The site, by the roadside overlooking Keem Strand at the western end of Achill Island, is a massive vein of quartz that occurs at the junction of some local quartzite and schist. It is possible that this vein was found during the construction of the road to Keem Strand, but what is certain is that this mineral locality had been noted as early as 1832 by Sir Charles Lewis Giesecke (1761–1833, professor of mineralogy at the Royal Dublin Society), and has been regularly reported and written about ever since.

Fabulous purple crystals of amethyst up to ten inches long were found in the nineteenth century, and many Achill locals subsequently collected the amethyst as a decorative mineral, and to sell to tourists. Unfortunately, in recent times, the site has suffered industrial-scale collecting, with people coming in with explosives and mechanical diggers. When I last went there in the 1990s, there was not a lot to be had. But Achill may still have treasures to offer. One local farmer apparently finds purple crystals as he ploughs his fields; the evidence suggests that a new, good-quality, amethyst locality might actually occur beneath his vegetable patch.

The second locality is Cork. A lesser known, but equally spectacular, amethyst locality occurred in Cork city, more specifically in the Blackrock area where large deep-purple amethyst crystals occurred in veins within the local limestone (see Plate Figure 8). Some really huge crystal masses were recovered, several feet across and with individual crystals several inches across. The site was discovered around 1770 and was worked, very much on the quiet, by a local jeweller. This suggests that jewellery was made from some of this material. The jeweller in question was doing very well. He had a small coterie of helpers who were sworn to secrecy but, unfortunately for him, he told his wife about the locality, and she went and gossiped. It wasn't too long before everyone knew about the site. They descended like vultures, and that was the end of that! The quarry where the crystals were found was eventually tarmacked over in the 1970s and became a coal merchant's yard. But one has to think: if there was one such locality, could there be another, as yet undiscovered?

Beryl (Aquamarine): $Be_3Al_2Si_6O_{18}$

Aquamarine is the variety of beryl that is transparent, seawater-blue, as might be seen on a sunny Mediterranean day. It is a close relative (I'd say sister) of emerald. The Mourne Mountains (County Down) are the classic Irish locality for collecting aquamarines, which may be found lining the many holes and cavities (former gas pockets) in the Mourne granites. Modern specimens can be up to 3cm in length, but these are a shadow of what was collected during the nineteenth century; some aquamarines were known to be at least 12.5cm long and of good gemmy quality.

Many were found by Mourne quarryman-turned-mineral-dealer Patrick Doran. It does seem likely that some of these specimens were made into jewellery, but proving it would be something else. Stephen Moreton, one of the few modern authorities on Irish minerals, has collected gemmy aquamarines from Diamond Rock (Mourne Mountains) and related localities, but there is never quite enough uncracked, flawless crystal to make a good faceted gem. Beryl, as a mineral, is not too rare and can be found in Dublin, Wicklow, Waterford, Donegal and Down. But only in Down is there real hope of gem-quality material.

Diamond: C

Ireland's one and only diamond is a controversial affair. It is known today as the Brookeborough Diamond. This stone, a slightly yellowish diamond, was initially, and allegedly, found in 1816 by a little girl who was bathing near the White Bridge in the Colebrooke River, County Fermanagh. The diamond was subsequently given to the owner of nearby Colebrooke House and estate, Henry Brooke, who had it faceted and mounted in a ring made of Wicklow gold. And that ring is owned by the Brookeborough family to this day.

The stone in the Brookeborough ring is a genuine diamond. Far less certain is the story of its discovery and whether it really came from the Colebrooke River. The local geology is that of a thick layer of glacial till on top of Carboniferous Period limestones. Not promising material. However, to be very generous about it, there are some nearby Devonian Period sandstones and conglomerates that contain, at least in part, metamorphic clasts within them. These clasts might have been ultimately derived from the even older Neoproterozoic-to-Cambrian metamorphic rocks known as the Dalradian Group, which dominate Donegal, Tyrone and Derry.

A possible, if unlikely, scenario might be that these old Dalradian rocks harboured some diamond-bearing kimberlites or lamproites (rock types in which diamonds tend to be found) and that these were eroded, with a few diamonds, to form the later Devonian sandstones of the Fintona Block. Then, very much later still, the Ice Age may have eroded these sandstones (and a few diamonds) to form part of the glacial till from where, at the very last geological instant, the Colebrooke River eroded the glacial till, and a diamond, which by this point would have had a long and eventful history over the previous 600 million years, popped out, was washed downstream, and was picked up by a little girl.

This unlikely scenario has, nevertheless, been taken very seriously. Mining and mineral exploration companies (such as Cambridge Mineral Resources from Bristol, England, and Poplar Resources from Vancouver, Canada) have taken out licences in the area to try to find either actual diamonds or some of the minerals that are associated with diamonds, termed 'indicator minerals'. Despite much panning, hammering and some drilling, no diamonds have yet been found. But, it must be said, there is just enough tantalising evidence from the presence of some indicator minerals to suggest that there is still a remote chance of striking it lucky. Even personnel from the Geological Survey of Northern Ireland, such as Ian Legg, have said that we couldn't rule out diamonds being present somewhere in the nine counties of Ulster.

Nevertheless, there is an understandable suspicion among professional jewellers and mineralogists that the Brookeborough

Diamond was planted. The problem is that there is no good evidence for any of the explanations. It might be a genuine find (in the sense that the diamond occurred naturally where it was found), or it might be a big old-fashioned fraud, albeit using a real diamond. Perhaps one line of enquiry, given that the geological leads have gone cold for the moment, might be to investigate the historical movements of the adult members of the Brookeborough family at the time: had anyone gone abroad to a potential diamond locality and brought one back? Did they know anybody connected with diamonds? And did they then get the little girl to play along with the ruse? The mystery of Ireland's only diamond goes much deeper than the Colebrooke River itself.

Garnet (Carbuncles)

As a mineral, garnet is not rare. It may be found in all manner of rocks in and around granites and in areas of what were once sedimentary rocks that have suffered (metamorphic petrologists like to say 'enjoyed') heating and squeezing that has turned them into metamorphic rocks. But gem-grade garnet (i.e. that could be faceted, polished and worn as fine jewellery) in Ireland is hard to find. Several nineteenth-century reports of gemmy garnets from various parts of Ireland have been personally investigated by Stephen Moreton, and he has found nothing there to interest a gemologist or jeweller.

However, gem-quality garnets, or carbuncles as they were once called, are known to have been worn by the ancient Irish, and some can be seen in the National Museum of Ireland – Archaeology. Where did the ancient Irish get them? Were the garnets native, or were they imported?

Malachite: $Cu_2^{2+}(CO_3)(OH)_2$

There is some evidence that malachite, the gorgeous green, often stripy, copper mineral, has been used in the past to make ornaments. Malachite in Ireland is moderately widespread, but good material for ornamental use is uncommon. Probably the best stuff came out of the nineteenth-century copper mine at Coosheen, County Cork, where chunks of some 50lb were recovered. Some of these were apparently carved into ornaments. Identifying any of these now would prove a challenge. The only other significant source of quality malachite was from Tynagh mine, County Galway. A few nice pieces can even be collected to this day. But, to the dismay of many, the vast majority of high-quality malachites from Tynagh sadly ended up in the crushers.

Pearls

Gem-quality pearls in Ireland come from the freshwater pearl mussel, *Margaritifera margaritifera,* and have been used in native Irish jewellery for well over a thousand years (see *The Irish Pearl: A Cultural, Social and Economic History* by John Lucey). For centuries, freshwater pearls have been fished from rivers and lakes from Donegal to Wexford to Kerry. And several examples of necklaces made from Irish freshwater pearls are still extant.

Given the importance and longevity of the native Irish pearl industry – which has all but died out now – it is no surprise that the Irish name for pearls, or for gems that are likely to be pearls, such as *némann* or *séd,* can be found in some place names dotted about the country. A selection of such names include the name of a tributary off the River Blackwater at Lismore, County Waterford, the Ownashade (*Abhainn-na-sed,* 'river of the pearl/jewel'); Loughnashade ('lake of the jewels') in County Roscommon; and from *némann* we get Lough Namona in County Kerry and Lough Naminn in County Donegal.

From the cruel twists of historical fate, notably the Viking raids and the destruction of monasteries by Henry VIII among many others, most of our ancient Irish gems, jewellery and gem-rich regalia have vanished. But we do get glimpses of what must have been many spectacular pieces; for example, we still have the mitre of Conor O'Dea, Bishop of Limerick (1400–1426), which can be seen in the Hunt Museum, Limerick city. This ornately decorated mitre, dated at about 1418, features many pearls.

Quartz (Rock Crystal, Citrine, Rose, Smokey): SiO_2

Quartz is silica, one of the most common minerals in the Earth's crust. You rarely have to travel too far in order to find some – which is not the same as finding good crystals of it. There are a multitude of places in Ireland where it is possible to find beautiful, if small, water-clear rock crystal. The most general piece of advice is to go to a place where the bedrock is made of a quartz-rich rock, such as granite or sandstone or quartzite, and look for where that rock has been ripped and torn (by geological forces) to produce cracks; there is a good chance that in the cracks in such places, quartz will have crystallised.

Many people will have heard of the 'Kerry Diamonds': ethereally clear quartz crystals found in cracks within the lightly metamorphosed Devonian sandstones (the Old Red Sandstone) of the Cork and Kerry region. These definitely have the potential for jewellery.

Where there is a rule, it is made to be broken. Although silica-rich rocks can yield lovely quartz crystals, so too can cracked and deformed calcium carbonate limestones (although calcite will be the main mineral in such cracks). But you can find quartz crystals up to 6cm long in deformed limestones if you are lucky.

One particular account must be highlighted here, simply because it is so remarkable. A short note, written in 1833, contains mention of absolutely huge, 70lb and upwards, quartz crystals at Benbradagh, near Dungiven, County Derry. What is more, apparently many comparable crystals could be found loose in the soil. How did this come about, and where are these huge crystals now?

In addition to the 1833 account, but on a less grand scale, even reports from 1829 (*Topographical Dictionary of the United Kingdom* by Benjamin Pitts Capper) tell of 'Dungiven diamonds' that '*when cut* have a very fine lustre' [my italics]. It seems these quartzes were used as faceted stones, but do any examples still exist?

True, but pale, citrine has been found once, and relatively recently, by a local collector somewhere in the Barnesmore Mountains of Donegal. Rose quartz, not to be confused with ordinary quartz stained with rust (iron oxides), is rare in Ireland, but has been reported from County Donegal: Bradlieve Mountain, Pollakeeran Hill and Maghery. At this latter locality, Stephen Moreton found pieces on the shore to the north of the strand that he says could be good enough for cabachon work (i.e. producing small polished domes for jewellery). But given that true rose quartz is rare in Ireland, I would leave it 'au naturel'.

Smokey quartz is known in Ireland from the early nineteenth century at least, and most famously from the Mourne Mountains (here with beryl and topaz). Some excellent crystals can still be collected from vugs and pegmatites in the granites here, and beginning collectors are recommended to put on their hiking boots and enjoy themselves in the Mourne Mountains. Smokey quartz of collectable quality is also to be found in the Barnesmore and Blue Stack Mountains of Donegal.

Sapphire (Corundum): Al_2O_3

We *do* have sapphires in Ireland. We *don't* have any of gem quality, or much bigger than a pinhead. In their hunt for diamonds (see above), the exploration companies that were scouring and test-drilling Donegal did find tiny sapphires in samples of stream sediment. Additionally, relatively large (note *relatively*) deep-blue sapphires up to half a millimetre in size have been found close to the Brockley dolerite plugs on Rathlin Island, County Antrim. Very small sapphire finds have been reported in other counties – notably in Wicklow, around Croghan Mountain; also in the metamorphic rocks surrounding the Omey Granite in County Galway; and at Tievebulliagh in County Antrim – but nothing for a gemologist to get excited about. Mind you, if I found even a tiny sapphire in a stream sediment sample, it would get my pulse racing.

Spodumene (Kunzite): $LiAlSi_2O_6$

Spodumene. First, and for no other reason than pure pleasure, let your tongue roll over the syllables … *spod'u'mene*. Actually speak it, slowly and with feeling – one can imagine comedian Rowan Atkinson enjoying this one. What a great name! Anyway, spodumene is a source for the metal lithium, vital for modern technology (e.g. batteries) and (ultimately) in treating depression in the form of Prozac. Kunzite and hiddenite are two of the gem varieties, being transparent pink and green, respectively. Leinster, in particular, hosts spodumene, most of it white or grey, cracked and showing various signs of having geologically suffered in several ways. But there is one report by Steiger and von Knorring (1974) of 'gemmy … kunzite, the pink gem variety' from Seskinnamadra, County Carlow.

Taaffeite: $Mg_3Al_8BeO_{16}$

Taaffeite is one of the rarest and most expensive gems in the world. Its type locality is in Ceylon; the nearest *in situ* specimens to Ireland are in Austria; yet it will forever be linked to Dublin. It was named after Count Edward Charles Richard Taaffe (1898–1967), a 'brilliant and unorthodox' gemologist who was born in Bohemia, the only son of Henry, Count Taaffe, the twelfth Viscount of Corran, Baron of Ballymote, County Sligo, and his wife, Magda, Countess Taaffe. The story of taaffeite's discovery is one of the most remarkable in all gemology.

In October 1945, Edward Taaffe was on one of his hunts around Dublin for gems when he entered the Fleet Street shop of watchmaker and fellow jeweller Robert Dobbie. Dobbie had a number of unsorted boxes of random gems and very many pieces of cut coloured glass (termed 'paste' in the trade), all of which had accumulated over the previous twenty years from when Dobbie's own father had run the shop.

Dobbie told Taaffe he could go through all the boxes, pick out the real gems and make him an offer. After the best part of three days sorting gem from paste, Taaffe paid £14 for what he had selected, noting that it was easy to do business with Dobbie. Back home, as he washed, sorted and tested all his purchases, one small, faceted mauve stone of 1.419 carats troubled him. It looked like a spinel (a group of hard metal-oxide minerals that come in many colours and are often used in jewellery), but it showed one very different sort of optical characteristic that spinel does not have – double refraction. Taaffe noted, 'In a certain direction every speck of dust on the back and every scratch appeared double like on a badly wobbled snapshot.'

On 1 November 1945, he sent the stone for analysis to B.W. Anderson at the laboratory of the London Chamber of Commerce, but Anderson was also unsure of what it was. Finally, in 1951, Anderson and some colleagues determined that it was not spinel but a brand-new mineral species; however, the testing procedures were physically destructive and caused the specimen to reduce in weight down to 0.56 carat (one-third of its original size). As well as being new, this mineral could boast two other firsts: it was the first new mineral species to be identified from having been an already faceted gem; second, it was the first gem to contain both beryllium (Be) and magnesium (Mg). Anderson named it 'taaffeite' in honour of our unorthodox count.

Topaz: $Al_2SiO_4(F,OH)_2$

The most prized topaz crystals are those that are a transparent golden sherry colour, with blue topaz coming in second. Ireland has one relatively well-known area for topaz crystals: the Mourne Mountains, around Slieve Binnian, Slieve Corragh and at Diamond Rocks. All the topaz crystals can be found in cavities in the host granite where the crystals have had space to grow unfettered. And some of them, though not of large size, are of great quality. They are almost all colourless, and I have never heard of any of them being made into jewellery. Stephen Moreton tells me, however, that a friend of his found 'a terrific gemmy yellowish crystal ~25 x 23 x 20mm in a vug in an old quarry on Slieve Binnian'. The potential for gem-grade material is still there, it seems.

Tourmaline

Tourmaline is a popular mineral for jewellery and for mineral collectors the world over because of its wide range of transparent to translucent colours. Some of the most prized tourmaline crystals have several colours in a zoned pattern (like a mouth-watering naturally crystallised fruit jelly), but this particular type is not found in Ireland. Tourmaline is quite common in Ireland, being primarily associated with all the major granites and with their associated metamorphosed chemical sediments (coticules and related rocks), but is normally jet-black and not of gem quality.

However, the mineralogy authors Greg and Lettsom (1858) tell of one find by Patrick Doran, that wily nineteenth-century collector who claimed to have found gemmy red tourmaline (the rubellite variety) from the Ox Mountains. Nobody in modern times has replicated that find. Nevertheless, given the amount of tourmaline about, one day someone will surely come across even a few crystals of indisputable gemmy material.

Turquoise: $Cu^{2+}Al_6(PO_4)_4(OH)_8.4H_2O$

Turquoise occurs in three places in Ireland, two of them discovered by Stephen Moreton. The first is at Ballycormick, near Shanagolden, County Limerick, where it forms thin coatings to vuggy quartz, which itself occurs in a brecciated gossan (i.e. a metal deposit that has been altered and broken up by the action of hot waters). The second is at Grouse Lodge Quarry, which lies about 5km south of Shanagolden, the turquoise here forming bluish-green coatings on joints and fractures that affect Upper Carboniferous phosphatic black shales. The third, and probably the best, is at the base of the coastal cliffs in a small bay northwest of Doon East, 1km north of Ballybunion, County Kerry, where turquoise forms patches and coatings up to 2 x 10cm across and 1mm thick. None of these occurrences produce turquoise suitable for lapidary use, however.

So ... Are there Emeralds in the Isle?

Ireland is universally known as the Emerald Isle. Shamrock-green is the colour associated with St Patrick's Day. Green is part of Ireland's national flag and the national team colours, be it for football, rugby, cycling or whatever. But have green emeralds ever been found in Ireland? True emeralds are a deep-green variety of the mineral beryl, the green colour being caused by small additions of the element vanadium, or sometimes chromium. Ireland does have crystals of beryl, including the pale-blue variety, aquamarine (see above), and, if one knows where to look, non-gem-quality beryl is not too rare. But emeralds?

There was once a rumour of a rumour from one of Trinity's geology lecturers that a geology student had found an emerald in a vein (a so-called pegmatite vein) in the local granite of the Dublin Mountains. But there is absolutely no evidence to support this. Beryl does, however, occur in the Dublin Mountains, and I've collected specimens myself. There are even popular books currently in circulation that state that emeralds have been found in Ireland. However, dear reader of this 'popular book', to my knowledge at the time of writing, *no* emerald has *ever* been found in Ireland.

The 'emerald' of the Emerald Isle is the green grass, which, if truth be told, is a far more valuable national asset than any number of emeralds. Ireland's climate and topography, allied to quite a bit of human modification to the landscape over the millennia, produces remarkable green sward. But of remarkable green emeralds ... there are none.

12

The Most Controversial Rock in Ireland – Granite

Introduction

Ah granite. A gorgeous rock. A rock so … innocent-looking.

Granite has been used as a building material for thousands of years and is associated with all things hard, rugged and durable. Made of three minerals – quartz, feldspar and mica – it is one of the two archetypical igneous rocks (the other being basalt), and it is one of the most common and distinctive rock types on Earth. Granite intrusions can range in size from small, metre-sized blobs to truly gargantuan masses that are many hundreds of kilometres in length, such as occur in the Cordilleran mountain ranges of North and South America. It is one of the very few rock types that almost everyone can recognise.

When it comes to trying to understand what granite is and how granite originated, many books have been published, dozens of conferences and symposia have taken place, and literally thousands of scientific papers have been written over a period spanning three centuries. Granites have been studied for as long as geology has been a science. Many geologists have dedicated their entire lives to studying granite from almost every conceivable perspective. So, after the best part of 300 years of the most intense scientific scrutiny, the human race should know just about all there is to know about granite, right?

Wrong!

Granite's extreme commonness and familiarity is a Janus mask. The ultimate origin of granite is a currently impenetrable secret. Mankind has produced geniuses who have discovered general relativity, evolution, quantum mechanics, chaos theory and that bread can be sliced. But *nobody* understands the true nature and origin of granite – one of the commonest rocks there is. Questions posed by any seven-year-old rock hound, such as: 'Dad, where does granite come from?', or, 'How does granite form?', cannot yet be fully answered, even if Dad is a professor of igneous petrology.

Granite is not just Ireland's most controversial rock; it is the *world's* most controversial rock. Why? Because every single characteristic of granite, bar none – the constituent minerals, every type of mineral texture, the chemistry of the minerals, the chemistry of the rock as a whole, the isotope data for the minerals and for the rock as a whole, the geological relations between the granite and the rocks that surround it, the 3D shape of a granite intrusion, and every other imaginable aspect – has been, and to varying degrees continues to be, vigorously debated. Place three or more 'granitologists' in a room, set a stopwatch, and see how long it takes for the arguments to start.

Why is understanding granite important? Because granite forms the key link in the chain by which we can fully know the origin and workings of the entire Earth's continental crust: from the development and evolution of large accumulations of sediment (so-called sedimentary basins – which can turn into granite upon heating, melting and squashing); the opening and closing of ancient oceans (on the margins of which very many granites are found); many metamorphic processes (the influence of hot granite on its surrounding rocks); the influence of hot fluids in, on and passing through rocks (granites, being a heat source, induce large-scale convection of hot water within the crust); the development of mountains (granites often form the cores of mountain belts); the development and siting of metal deposits, so crucial for the modern way of life (granites heat water in the crust that, in turn, scavenge metals from the rocks intruded by the granite and these dissolved metals can be redeposited in concentrated form in fracture veins); the influence of the Earth's mantle on the continental crust (possibly telling of the underlying mechanisms in granite formation): i.e. pretty much everything of significance in many branches of geological science.

When we *fully* understand granites, we will have a profound knowledge of how the Earth's continental crust formed, how it has evolved and how it will continue to evolve. Granite is the rock that links all these things together. Scientifically, the stakes could hardly be higher.

What of Irish granites in particular? The granites of Ireland have played key roles throughout humanity's long quest to understand granite. And what is more, they continue to play a key role.

This essay is, like them all, written for everyone, but perhaps it is the one in this book that might give the professionals something to chew on. There are profound geological questions still to be addressed. And further study of Irish granites could help address them. But first, let's look at our truly world-class granite heritage.

Irish Granites in Geological History

The Leinster Granite

The Leinster Granite is one of the longest-studied granites in the world. Scientific papers relating to this granite, or in which the granite features very prominently, date back to the early nineteenth century. One of those was published in 1812 by William Henry Fitton (1780–1861) and is itself based almost entirely on even earlier observations made by the Reverend Walter Stephens, who had studied the granite in the late eighteenth/very early nineteenth century, but died in 1808 before publishing his findings. Fitton published them on Stephens's behalf. Not long after, in 1819, English mining engineer Thomas Weaver (1773–1855) published a much-cited account of the geology of southeast Ireland in which the Leinster Granite featured large.

New minerals were discovered from this granite and its associated rocks at an early stage. In 1818, the new mineral 'killinite' (see Chapter 10) was described, it being named after Killiney in south Dublin. And the mineral 'haughtonite' was also put forward

Plate Figure 1. The Irish flag – Nature's version (County Dublin).

Plate Figure 2. The Blue Waterfall (County Waterford).

Plate Figure 3. World-class folds at Loughshinny (north County Dublin).

Plate Figure 4. Above: Great Sugar Loaf (County Wicklow).
Below: Conchagua stratovolcano, El Salvador.

Plate Figure 5. Top left: thigh bone of *Scelidosaurus*. Top right: *Scelidosaurus*. Bottom left: thigh bone of *Megalosaurus*. Bottom right: *Megalosaurus*. (Bones from County Antrim.)

Plate Figure 6. The author and Irish gold (counties Dublin and Wicklow).

Plate Figure 7. Cotterite (County Cork).

Plate Figure 8. Huge amethyst group from Cork.

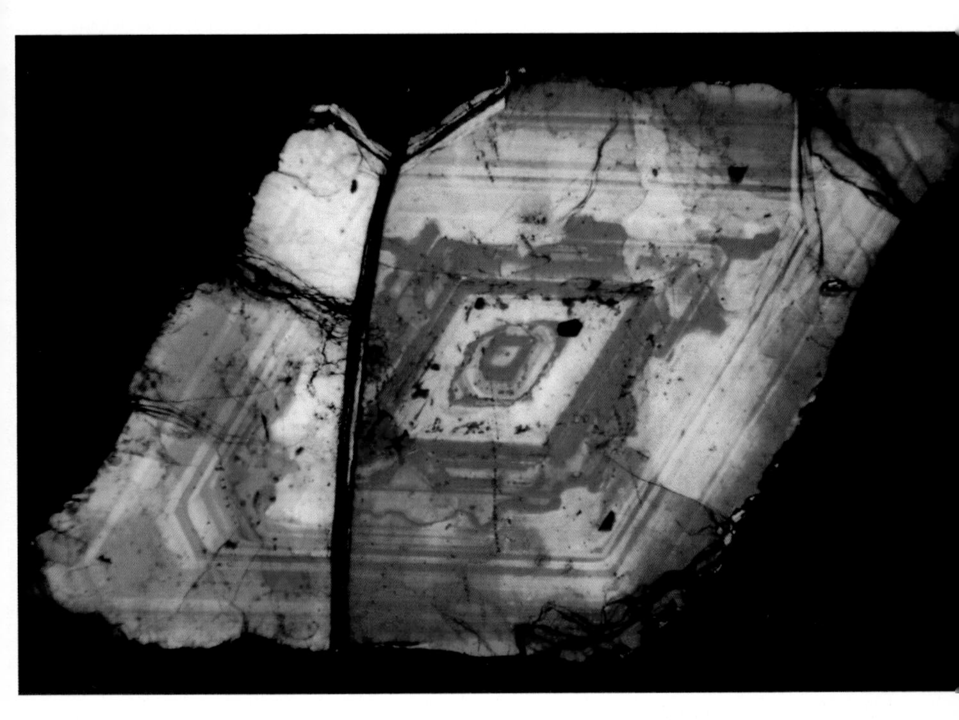

Plate Figure 9. Zoned muscovite (above) and sparkling granite (County Dublin).

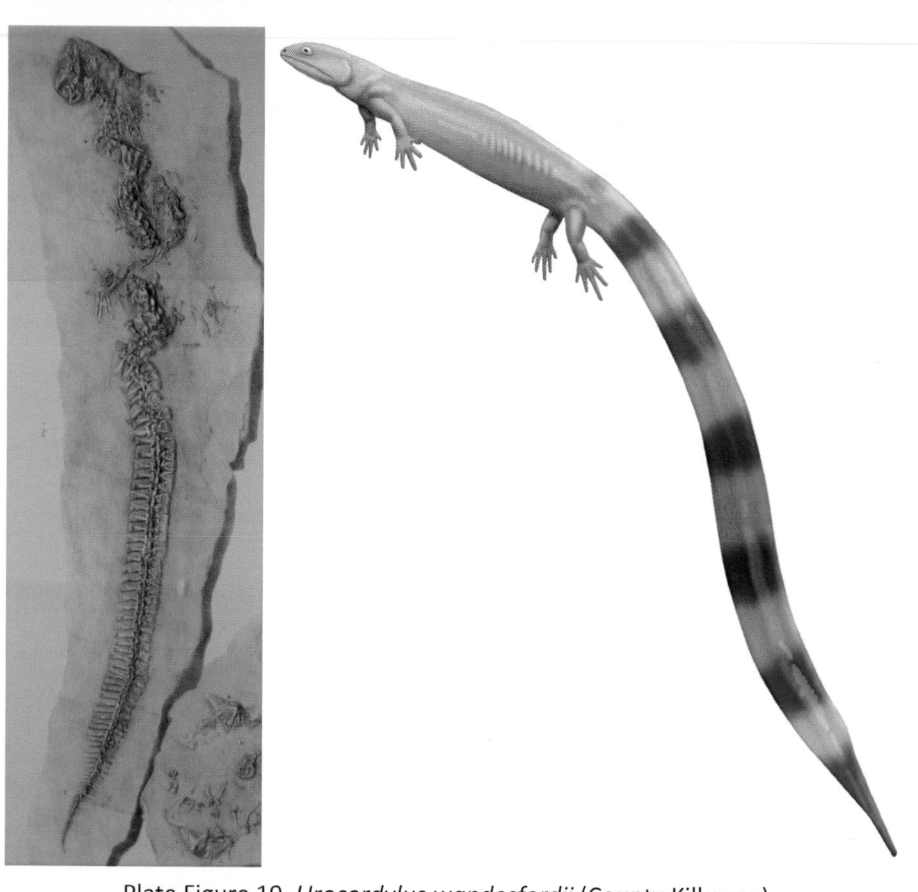

Plate Figure 10. *Urocordylus wandesfordii* (County Kilkenny).

Plate Figure 11. Giant Irish deer (County Limerick).

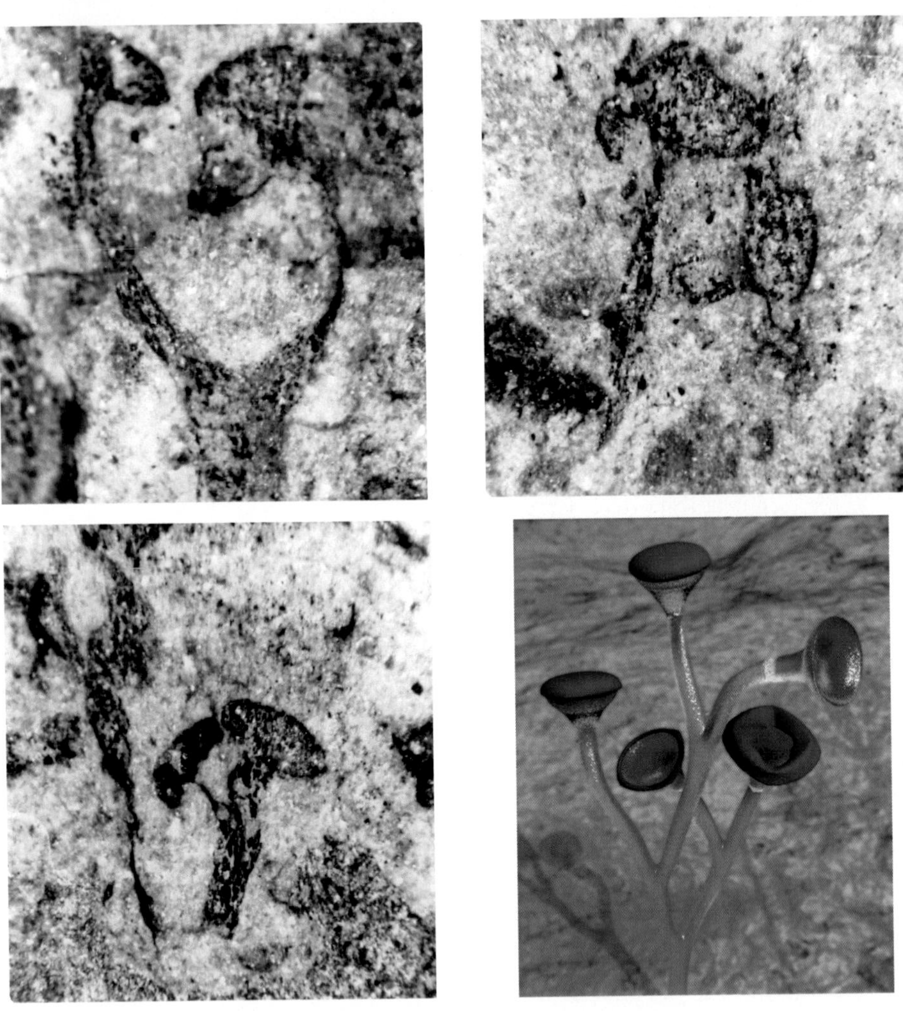

Plate Figure 12. *Cooksonia* (County Tipperary).

Plate Figure 13. *Siphonophyllia samsonensis* (County Sligo).

as a new variety of biotite, it being named after Reverend Samuel Haughton, professor of geology at Trinity College Dublin. Both minerals featured prominently in nineteenth-century mineralogy books, academic papers and popular articles, not just in Ireland but internationally.

In 1856, Samuel Haughton became one of the first people to perform quantitative chemical analyses on granite, i.e. putting numbers to the amounts of different chemical constituents, not merely saying that the granite contained them. Indeed, William Sollas, Trinity College Dublin professor of geology and mineralogy, claimed, in 1891, that Haughton's analyses were the *very first* quantitative analyses of *any* igneous rock in the world, no less. Today, there are literally hundreds of thousands, possibly millions, of such analyses now published, with hundreds more every year. This now-standard procedure may trace back to Ireland.

John Joly, the Trinity College Dublin professor of geology and mineralogy who succeeded Sollas, did pioneering work on radioactivity and absolute ages of rocks while studying the Leinster Granite. Granite contains the mineral biotite, a dark-brown variety of mica. When the biotite is sliced down to a very thin sheet, dark haloes can often be seen that occur around really tiny minerals that had got included into the biotite as it grew. This feature had been seen many times in many granites, but nobody knew what these odd haloes were.

In 1912/13, Joly published an article in which he examined biotites from some Carlow granite and demonstrated that the tiny mineral inclusions within the biotite were radioactive; that the haloes were due to the radioactive decay of different elements (and different isotopes of different elements); that it indicated that the biotites, and their host granite, 'are old, appallingly old'; that the visible expression of these haloes is of radioactively damaged host-biotite crystals; and that physics calculations on such haloes can tell of their (great) age. This was very influential work.

All this research on the Leinster Granite found its way into international books, university texts and made the Leinster Granite, especially during the entire nineteenth century and the early part of the twentieth century, a granite that geology students in the English-speaking world had to know something about.

The Donegal Granites

If the Leinster Granite was one of the granite 'stars' of the nineteenth century, then the Donegal granites became the global superstars of the second half of the twentieth century. Today, they are regarded as 'classic'. The reason for their superstar status is the combination of their intrinsic complexity that tests the very best geological minds; their being subject to intense study for many decades; their being used as classic examples of various types of emplacement mechanisms, i.e. the proposed different ways by which the granites move from their place of origin to where they come to 'rest' in the crust to cool and crystallise (the main reason for their fame - see below); and the fact that a series of very eminent international geologists chose to study them and base their careers on them. The most eminent of these was the late professor of geology at Liverpool University, Wallace Spencer Pitcher (1919–2004), known to all as Wally Pitcher.

Pitcher was a London-born man who, in the late 1940s, chose to focus on the Donegal granites as the research topic for his PhD, under the supervision of the legendary granite man of his generation, Professor Herbert Harold Read (1889–1970). Pitcher and his colleagues found that the different granite units (termed 'plutons') comprising the entire Donegal granite mass (termed a 'batholith', which is a mass of granite formed from individual plutons) had been emplaced – got to where we see them today – in a variety of characteristic ways.

For example, the Ardara Granite seems to have formed via an expanding ballooning dome mechanism, i.e. inflating like a balloon, only being pumped full of semi-molten granite rather than air; the Rosses Granite seems to have been formed via the subsiding of sharply defined cooled blocks of already formed granite that has gently subsided into a softer underlying granite sheet (a cauldron subsidence mechanism); and the Main Donegal Granite seems to be a thick tongue of granite that inserted and probed its way into the surrounding host metamorphic rocks as a series of almost vertical sheets, the multiple tonguing making its way in as the surrounding metamorphic rocks were being gently pulled aside due to an opening shear zone.

These various emplacement-mechanism ideas of ballooning, subsiding blocks and sheeted tongues played an immense role in interpreting other granites from all over the world. The entire set of South American granite batholiths, for example, has been explained using these ideas (see also Mourne granites below), and these mechanisms continue to dominate geological thinking to this day, with diagrams and references to one or other of the Donegal plutons commonplace in textbooks the world over. In the Donegal batholith, we have a real international geological icon that has influenced countless articles on countless other granites.

Before leaving Donegal, there is one other classic granite feature that must be mentioned: a spectacular variety of granite called an orbicular granite, which can be seen on the coast at Mullaghderg. While not the very first example of an orbicular granite ever studied (that was in Poland), it is, nevertheless, one of the earliest (described by Frederick Henry Hatch in 1888). Orbicular granites comprise rude spheres of granite or granitic minerals that seem to either float in or be packed into what looks like normal granite background rock. They are really odd things. And the one at Mullaghderg is odder than most because, in one part of it, there is a large granite 'sack' (like a really big orb, many feet across) that contains within it many hundreds of smaller granite orbs. Nobody still has any real idea what is going on here.

The Mourne Granites

The Mourne granites have achieved world fame ... and world infamy. Let's deal with the fame first.

James Ernest Richey (1886–1968) was the Tyrone-born son of a clergyman who received his school and college education in Dublin. He was precocious, highly enthusiastic and very inspired by Professor John Joly, who had taught him geology at Trinity College Dublin. Because of his passion for fieldwork, he abandoned a burgeoning academic career and went off to join the Geological Survey of Scotland. His enthusiasm was such that on his holidays in 1925 – freshly married to Henrietta, and with her in tow – he mapped the entire Mourne Mountains in six weeks. This would have been a prodigious feat in its own right but, during the process, Richey also identified a then-new mechanism of granite intrusion (i.e. an emplacement mechanism) called 'cauldron subsidence and ring-dyke emplacement'. This is where a large block of country rock (rock that the granite is intruding) sinks along outward-dipping, ring- or cone-shaped fractures, and the granite wells up into these large fractures to fill the space made available. Thus, the granite that comes between the top surface of the sinking block and the roof of the unmoved host country rocks above (remember – this is all happening underground) forms what we now see as the granite plutons: the Mourne granites. Richey invented the 'G' terminology to label different granite types, and the order of their intrusion (i.e. G1, G2, G3, etc.), something that is used to this very day. All this on a six-week holiday, effectively on his honeymoon! The idea of ring dykes and cauldron subsidence of large blocks has been applied to many granites the world over, and remains a powerful idea today.

And now the infamy.

A gargantuan academic spat raged for quite some time between the 1940s and the 1960s; indeed, it still rumbles on in some quarters. The spat centred on whether granites are really igneous rocks or whether they are some sort of metamorphic rock. The troublesome concept behind this particularly acrimonious period in granite studies was 'granitisation'. The essential idea here is that granites are not igneous rocks (formed from molten magma) in the way that basalts or gabbros are, but that they are formed by some kind of chemical transformation process in which elements such as silicon, sodium and calcium have migrated in, and/or elements such as iron, magnesium, potassium and aluminium have migrated out. These postulated waves of chemical migration were termed acid fronts and basic fronts, respectively. In this way, bodies of granite could be made more or less *in situ* just by mass transference of elements, and it would explain the more 'basic' rocks that, observation attests, usually lie outside a granite's margins. A clever idea, but extremely divisive for geologists of the time.

The originator of this idea was an exceptionally strong-minded woman called Doris Livesey Reynolds (1899–1985), the second wife of the venerable geologist Arthur Holmes. Reynolds came up with the idea of granitisation while studying the Newry granodiorite between 1943 and 1947. Women in geology at this time were rare: Reynolds was unique. Very forthright in her opinions, she once attended a lecture by a man who was advocating the magmatic view of granites and, from the back of the darkened lecture room, she loudly commented, 'Bollocks!' And this was in the 1950s in a room full of men. You didn't mess with Doris Reynolds.

Today, the idea of granitisation is only mentioned in 'history of geology' courses. Almost nobody believes in it nowadays. Nevertheless, it did force people to marshal their evidence for granites being in some way crystallised from a liquid magma and for everyone to more closely examine their own rocks. It was a great catalyst for new research. But the concept of granitisation – one of the most infamous ideas in geology – started in the Mourne granites.

Patterns, Patterns, Everywhere // But What Does It All Mean?

The following stanza is as true today as it was when it was written by the anonymous, miserable wretch driven to near madness in his attempt to understand granite:

The Rime of the Gryzzled Granitologyste (excerpt)

Patterns, patterns, everywhere,
Defying contemplation;
Patterns, patterns, everywhere,
Nor any explanation.

Where there are patterns in nature there are truths to be uncovered. Finding these truths may be difficult, but patterns always indicate that there is a 'deep truth' to be revealed. In the case of the Leinster Granite, we see many patterns. The problems arise when one superimposes the patterns from different sets of data onto one another to try to make sense of the whole lot together. In the normal course of events, everything should slip effortlessly into place. But granite is not a normal rock, and one is faced with a fiendishly difficult problem of reconciliation.

Two Popular Ideas on the Origin of Granite – But Are They Correct?

There are many currently popular ideas on the origin of granite. But some involve processes that seem inapplicable to the Leinster

Granite, as discussed below. One such process is large-scale chemical fractionation of a liquid granite magma. An analogy here is with the process of crude-oil fractionation. Start with dense crude oil and then separate off the heaviest part. This leaves a slightly less heavy residue that itself can then be separated into a relatively heavier and lighter fraction, which itself can be separated into a relatively heavier and lighter fraction, and so on, until one ends up with very light petrol, or even gas, at the end of the process. Thus, one has separated (fractionated) the original crude oil into its component parts, ranging from the heaviest molecules (tars and bitumens) to the lightest (kerosene and gases).

Geologists often assume that granites can do the same sort of thing and that this can explain the different granite varieties that can often be mapped out. In this scenario, the granite is essentially liquid at the start, then it starts to crystallise and thereby causes crystals (solids) to separate out. The remaining liquid part rises or separates, then some more crystals separate out from that now less dense liquid, and this residual liquid itself now produces crystals. One is left with a lighter liquid, and so on, until one has fractionated the original magma into its heavy to lightest components, and developed a series of sequential granitic rocks in the process. These different fractions are taken to be the different granite rock types we see in an outcrop and the different fractions have produced some of the chemical trends measurable when granite samples are analysed back in the lab.

Fine in theory. Unfortunately, this mechanism predicts a reasonably homogeneous distribution of certain isotopes of certain elements across all the different granite rock types because isotopes of elements cannot be chemically fractionated that much by such a mechanism. If fractionation was a significant process in forming the Leinster Granite, then we should see a similar, fairly uniform distribution of various isotopes. That is exactly what we do *not* see.

Another very popular mechanism currently in vogue in explaining how a granite gets to be where we now see it in the crust is that the granite magma was produced in the lower crust, some 30km or more beneath the surface, and this molten magma rises upwards through dykes and faults in the deep crust, as described in a very influential study by Clemens and Mawer (1992), like a kind of upward-directed peristalsis swallowing motion. This 'upward swallow' (my term) transfers the melt from the lower crust up into the higher crust where it ponds, is intruded by more melt coming up from below, and mixes.

Again, fine in theory. But such a mechanism should result in quite a random, or certainly very mixed, set of isotope analyses for any set of samples taken from the whole granite: after all, why would we expect any coherent patterns on either a local or, even more so, on a regional scale (e.g. the whole of Ireland) if the sources of the granites were 30km away or more, beneath where the granite arrived, the granite having travelled up as very many separate blobs and pulses, and over a wide area, presumably from a complicated mix of melting and faulting conditions? How such a mechanism can account for the fact that Leinster Granite isotope values are *not* random (see below) but preserve patterns on both a local (metre) scale and a batholith-wide (tens of kilometres) scale is very difficult to envisage.

Patterns in the Leinster Granite – and Beyond!

Now let's have a look at some of the patterns that occur across the whole of the Leinster Granite (i.e. the batholith) – and, yes, beyond.

The Map Patterns of the Granite Varieties in the Leinster Batholith

Geologists tramp over hill and dale plotting on a base map the various varieties of granite that can be physically and visually distinguished in the field. The Leinster Granite batholith is made up of five plutons, each showing mapped patterns of one sort or another. The Northern Pluton and its western blister extension – the 'side-car' bulge of the Upper Liffey Valley Pluton – shows, to a first approximation, that the granite varieties map into three bulls-eye patterns: two centred on the northern and southern parts of the Northern Pluton, respectively, and one on the central part of the Upper Liffey Valley Pluton. All three have a crude mirror-image symmetry about them. Furthermore, one could draw a line down the mid-axis of the whole Northern Pluton, from Dún Laoghaire (County Dublin) to Corragh Mountain (about 12km due south of

Blessington, County Wicklow), and one could see a crude mirror image of the granite varieties either side of this line.

Mappable patterns in the rest of the batholith (the other three plutons) are not as obvious, but one can, if one trusts the mapping, pick out a few large-scale features. In the Lugnaquilla Pluton, which starts only a few dozen metres south of the Northern Pluton but is separated from it by a screen of local metamorphic rocks, the granites seem to have a very rough east to west distribution, looking like a series of NE–SW bands rather than the bullseye patterns of its northerly pluton neighbours. The Tullow Lowlands Pluton has been very eroded and, much (much) later, buried under glacial sediments. Its shape, when viewed on a map, is similar to that of the continent of South America: wide at the top and tapering down to a point in the south before it meets the southernmost granites of the Blackstairs Pluton.

Now, this next statement must be taken with some caution because the quality of mapping of the Tullow Lowlands Pluton is, perforce, not as detailed as the rest of the batholith, but one might see a sort of large-scale east to west banding of the different granites, very crudely analogous to that of the Lugnaquilla Pluton. The southernmost Blackstairs Pluton has, to a first approximation, a rough NE–SW series of bands, too.

Thus, there are patterns discernible when mapping out the granite varieties in all five of the plutons of the Leinster batholith, although by far the most obvious are those of the two northernmost plutons.

Almost All the Leinster Granites Contain Shiny, Silvery Muscovite

Muscovite is a mineral that light can pass through, be reflected by or be refracted by. One of the properties possessed by many minerals is that they can bend polarised light, and, in the case of muscovite, the amount of bending can be correlated to its chemical composition. Way back in 1856, Samuel Haughton (1856b) made the observation that the muscovites of the Leinster Granite refract polarised light more as one goes north to south down the entire length of the batholith. That is, the muscovites at the northern end bend light to a relatively low degree, while the muscovites further south bend light relatively greater. Haughton did not correlate this observation to muscovite chemistry, but now we can say that this means there is a rough, but systematic, trend in muscovite purity as one goes south and, conversely, that muscovites in the granites get chemically more impure as one goes north.

The General Composition of the Granites

The great Samuel Haughton (1856c) again made some interesting observations based on chemical analyses of the granite itself. All the observations made at this time are very broad-brush, of course. Nevertheless, he found that, as one goes north to south down the length of the batholith, the amount of silicon increases (the southern granites are richer in silicon, generally). He also found, going north to south, that there is a decrease in the amounts of iron, potassium and aluminium. Just over a century later, in 1972, the English geologist Anthony Hall proposed something similar in terms of what might be thought of as 'theoretical mineral proportions' based on spot granite chemical analyses: his proposition was that as one goes north to south, there is an increase in quartz, an increase in orthoclase (potassium) feldspar, and a decrease in albite (sodium) feldspar. I must note, however, that more modern work is needed on general Leinster Granite chemistry and that both the quoted papers here require modern firming up.

The Evidence from the Isotope Patterns

Peter Mohr was a fellow PhD student with me in University College Dublin, and we both had Dr Pádhraig Kennan as our supervisor. Peter found a remarkable pattern, and one that I remember got Pádhraig very excited, that showed up in the northern plutons from what is known as – and please don't run away! – the 'strontium initial ratios'. This is just a number derived from a graph. It acts as a pointer to what original rock types might have melted to produce the granite in the first place. In more detail, this is the ratio of two isotopes, i.e. the same chemical element but with different atomic weights, of strontium (symbol Sr; it's also the element that gives the red colour to fireworks) – namely, ^{87}Sr to ^{86}Sr – that tell what the ratio of the two isotopes of strontium was in the granite at the time of the granite's intrusion ... and this helps define the essential type of rock that melted to form the granite in the first place. Which means that this ratio tells us something about the actual source of the granite: what it was before it melted to become a granite.

When that ratio is relatively high, it signifies that the granite formed from melting a lot of continental crust (acid rocks including sedimentary sandstones, siltstones, etc.); when this ratio is relatively low, it means that the granite had an input into its formation from rocks that are in some way related to the Earth's mantle (basic rocks such as various types of basalt). Peter found that the margins of the northern plutons were more basic, while the centres were more acid, and this must reflect aspects of the granite's source rocks, i.e. what precursor rocks melted to form the granites.

The major pattern that Peter found was that the Sr initial ratios form a roughly symmetrical mirror-image pattern either side of a NE–SW line drawn down the central axis of the northern plutons, similar to that from the mapped granite varieties (see above). But there is one crucial difference. The isotope pattern did not correspond in detail to the pattern of the mapped granite varieties themselves. Isotope trends crossed over visibly different granite types; equally, isotope trends could be traced *within* what looked like just one granite type. In other words, where one might (from some current theories) expect different-looking granites to have their own isotopic signatures (fingerprints), Peter found that the isotope patterns do not correlate with what the granite looks like in the field, and that even the same-looking granite can show (cryptic) patterns across it, which you would never guess from just looking at it.

These isotope trends are definitely patterns and they are definitely there: there is no sense, either, of any randomness or of homogeneity. Furthermore, Peter backed up his strontium pattern by finding an almost identical trend using a different isotope system, that of samarium–neodymium (Sm–Nd). Very curious.

The Big Regional-scale Isotope Pattern

Pádhraig Kennan himself spent many years collating every single strontium isotopic analysis from every single Caldeonian-age granite (i.e. granites about 400 million years old, like Leinster) that had ever been analysed in Ireland, England, Wales and Scotland – everything: published and unpublished. He manually plotted these hundreds of analyses onto a simple graph and, in a 1997 Royal Dublin Society paper, he published a really remarkable pattern. On this graph, the strontium 'character' of all the granites fall into clearly defined, separate (non-overlapping), yet adjacent and parallel fields. He could distinguish the separate isotopic character of many different granites but place them as belonging to just four regions: Scotland; central England and the Saltee Islands (Ireland); the Lake District (England); and the granites from the County Longford/County Down catchment area (Ireland).

But because the isotope characteristics are both distinctly separate from one another, yet run on the graph parallel to each other and are adjacent to each other, the implication is that we are seeing a grand reflection of the sources of all these granites. The granites' sources could, on a regional scale, be some sort of coherent entity – distinguishable, related, but not mixed. To the casual reader, this last statement may produce a 'Hmm, interesting enough' reaction; some academic geologists might react with an explosion of expletives and spittle.

To suggest that all the many dozens of British and Irish granites – formed, wisdom would have it, from many different processes, from different source rocks, at different times, at different depths of emplacement and depths of origin (and so on) – might *somehow* be coherently related could cause one to choke on one's coffee.

Further choking will ensue when geologists examine the growing evidence for the origin of those thick veins of large minerals that are commonly seen to cut through the granite, veins of coarse crystals known as pegmatites. Most geologists consider pegmatites as being the residual liquid left over from the mass of granite crystallisation (as a type of fractionation product, if you remember our oil analogy above). But evidence is mounting that a great many pegmatites might really derive from liquids sucked out of the rocks that the granites intrude, i.e. they don't derive from the granite at all but from partially melted country rocks. But the facts are slowly gathering and starting to speak up, and they are not happy with the textbooks.

Cracking the Leinster Granite Nut Could Crack All Such Nuts

'Patterns, patterns, everywhere // But what does it all mean?' We can be certain that patterns in nature reflect some sort of underlying 'fundamental' that, if we can decipher it, will tell us a lot about how that aspect of nature works. I would love to be able to tell you, dear reader, what all the patterns in the Leinster Granite mean: what they are a reflection of. But I can't. And neither can anyone else. But let's look at a possibility.

Unlike just about every other granite in Ireland and Britain, the Leinster Granite now has several, independent, mutually exclusive patterns. It is likely that other granites also have them, but they have just not yet been identified. Without doubt, some of these patterns do need shoring up in order to be convincing. Granites are igneous rocks and have been produced by the melting of some pre-existing rocks: their source rocks. The source for the Leinster Granites, based on the chemistry, minerals and various isotopes measured from the granites, *suggests* that the Leinster Granite batholith was formed from the melting of a coherent mass of Lower Paleozoic sediments with some volcanic rocks in its original upper parts for added flavouring. The patterns suggest that the rocks that formed the granite at its southern end formed close to a continental margin, whereas the granite's source rocks at the northern end were from deeper-water sediments that had more volcanic rocks within them.

There has been no mixing or mingling of granite varieties to any great extent because if there had been this would, practically by definition, ruin the patterns. There has been no fractionation of a large body of liquid magma because the isotopic variations in the granite are too great for a chemical fractionation mechanism to explain. And could one really induce and preserve the observed patterns if one invokes the currently popular idea that granite liquids originate deep in the lower crust and this liquid, or mush, is pumped vertically or obliquely upwards through faults for 30km to be then injected one into another? You, the reader, are now at the cutting edge of granite studies thanks to studies on Irish granites. But it is a place where most geological angels fear to tread.

Irish granites could play a really significant role in the future of granite studies. Three things need to be done. Replicate some of the research done on Leinster itself (in order to be sure to be sure) and then also see if the results hold true for other granites elsewhere. Take the facts, as opposed to the myriad theories, as they are now and think about how to explain *all of them together*. Continue to research the Leinster Granite (including the various veins one sees with the granite, such as the coarse-grained pegmatites and the fine-grained aplites), and the rocks that surround the granite. The Leinster Granite nut is possibly the closest to cracking the whole enigmatic, endlessly frustrating, compellingly fascinating and scientifically vital granite problem. If we can find a full and convincing explanation for the origin of the Leinster Granite, then we effectively have it for the world.

13

Your County's Rock, Mineral, Fossil and Geo-nickname

Introduction

In this chapter, I am going to suggest a complete set of geological emblems for each county of Ireland. Each county is special, and although geology disregards human borders, every county has within it something that it can call its own.

Most Irish counties have an official coat of arms (except for North Tipperary and Armagh) that comprises a shield decorated with such things as dragons, castles, fountains, red hands, light-houses or wildlife. Most, but not all, also come with an official motto, either in Irish, Latin or English. However, there are no official Irish county flags. The colours we see on festive bunting, jerseys, and at Croke Park in Dublin are actually the Gaelic Athletic Association (GAA) county colours: these GAA colours have effectively become our proxy county flags.

Every county also has a nickname, sometimes several. Some of the nicknames are relatively old ('The Model County' for Wexford dates from 1847), while others were composed more recently ('The Faithful County' for Offaly dates from 1953). There are well-known examples such as Kerry 'The Kingdom'; Armagh 'The Orchard County'; Clare 'The Banner County'; and Louth 'The Wee County'. And we even have two geologically nicknamed counties. Kilkenny 'The Marble County' is named after the extensive quarrying of 'marble' (more correctly, limestone) and the use of this stone in many locally important buildings; and Monaghan 'The Drumlin County' is named after the great number of drumlins and its 'basket of eggs' topography. I decided to keep Monaghan's nickname because it is accurate. Kilkenny's was another matter: my geological conscience would not allow me to keep 'The Marble County' because Kilkenny has not one quoin of indigenous marble within it! But it is a good nickname, so I have used it for Galway instead, to celebrate the world-famous Connemara Marble. Note that Waterford 'The Crystal County' derives its nickname not from natural crystals but from its excellent glassware.

So, to all intents and purposes, every county today has a coat of arms, county colours and a nickname. I have taken this concept and applied it to the geological heritage of each county. Everyone can let the world (or at least the rest of Ireland) know what's special about them. Herein, I offer to each county its county rock, county mineral and county fossil, and all under a geologically appropriate county nickname. If you (the reader) have a designer streak in you, then have a go at designing a county flag for each rock, mineral or fossil suggested, or even combine all three into a magnificent triumvirate geo-coat-of-arms for your county. Submit it for official approval, why not? Fly the flag for geo-heraldry.

Criteria for Selecting Geo-Emblems

I applied the following principles to each county, while simultaneously considering the requirements of every other county, so as to avoid clashes or duplications. The emblem in question can or should:

- Be found in the county in question
- Be characteristic or distinctive of the county in some way
- Be either rare or common, as long as it accords with the above points
- Be uniquely chosen: not selected for another county, even if it might occur there also

When examining the geological literature to find the geo-characteristics for every county, certain challenges and problems arose. One odd problem was that some counties are very rich in material to choose from. It was not easy, for example, to settle on what rock might represent Wicklow or Mayo, what mineral might represent Antrim or Tipperary, or what fossil might be emblematic of Wexford or Kilkenny: all have multiple, and equally deserving, candidates. Equally, some counties are not as geologically diverse, either individually or from others: choosing a characteristic rock emblem for the many essentially all-limestone Midland counties, yet not choosing 'limestone' each time, was tricky. Many counties have similar fossils (and many have similar minerals) so one had to occasionally choose a fossil that was neither unique nor especially 'diagnostic' of that county, but that nevertheless did occur to a significant degree and did not repeat a fossil chosen for another county. Some decisions were obvious; others were a balancing act of weighing several factors simultaneously. But every county's need for individuality was taken into consideration.

The county geo-nickname usually relates to one of its county rocks, minerals or fossils. But not always. There are cases where I have chosen something quite different, and for a very good reason: there may be some overarching characteristic that, in my view, trumps any one of the individual characteristics.

Below is my selection of county rock, mineral, fossil and geo-nickname for all 32 counties of Ireland. Be proud of yours.

ULSTER

Antrim

The Basalt County

Antrim's most dominant rock type by far is the early Paleogene Period basalts, making basalt the characteristic rock of Antrim. Furthermore, it is this rock that forms the world-famous Giant's Causeway. And the Giant's Causeway is one of Ireland's only three UNESCO World Heritage sites (the others being Skellig Michael and Brú na Bóinne).

County Rock: Basalt

Antrim's bedrock geology is completely dominated by flow upon flow of basalt that erupted through dykes and fissures as so-called 'flood basalts'. These eruptions were due to the continental crust here being stretched apart; this was due to the even bigger process of the North Atlantic opening at the end of the Cretaceous Period and during the Paleocene Epoch of the Paleogene Period, an event that split Ireland from Greenland and Ireland from Newfoundland (we were all once neighbours!). During this time, Ireland got stretched and cracked and mantle basalts welled up through these cracks … but we didn't completely break apart. One of the natural wonders of the world is the Giant's Causeway, formed through local ponding of basalt and its subsequent cooling, cracking and solidifying, like a hot version of surface mud cracks extending downward.

County Mineral: Gmelinite

Gmelinite is a gorgeous, sorbet-orange to salmon-pink zeolite mineral that is the *only* Irish type-locality mineral (i.e. a mineral species or series defined from specimens taken from a particular locality where it was first found and characterised) that has the official rank of a whole mineral series. And the type area is Little Deer Park at Glenarm, Island Magee. There are three distinct varieties of this mineral, dominated by calcium (Ca), sodium (Na) or potassium (K) as the case may be. Although not a common mineral, it can be locally abundant, and it is well worth seeking out in the Island Magee peninsula.

County Fossil: Megalosaurus

Given that, at the time of writing, there have only ever been two dinosaur fossils found on the island of Ireland (see Chapter 7 and Plate Figure 5), one of them had to be a county fossil. And I have chosen the dramatic Jurassic meat-eater *Megalosaurus*, a bone of which was found on the beach at the Gobbins, Island Magee.

Armagh

The Shark County

A good and memorable county name, chosen because of the locally common (but not nearly so common elsewhere in Ireland) number of shark's-teeth fossils that can be found in the Carboniferous limestones near Armagh city.

County Rock: Granophyre

Geological disputes are always good fun … after the event and if one wasn't in the firing line being metaphorically shot at. Armagh saw a good one in the 1950s between the controversial and bullish Doris Reynolds (see Chapter 12) and a pair of formidable geologists called Edward Bailey and William McCallien. The dispute centred on the igneous rocks of Slieve Gullion: were they extrusive (poured out over the surface), which Reynolds wanted, or were they intrusive (intruded in other rocks but remained beneath the surface), as Bailey and McCallien wanted? The two lads won the day (more or less): Slieve Gullion is the innards of a large Paleocene Epoch volcano. One of the most disputed rocks in the area was granophyre, a granite-like rock dominated by a wormy quartz–feldspar intergrowth that we would now say is a mineral texture indicative of rapid crystallisation.

County Mineral: Feldspar

Possibly more so than other counties, Armagh shows great variety in its types of feldspar (a potassium–sodium–calcium aluminium silicate). The county rock of granophyre, at Slieve Gullion, is characterised by having its feldspar occur as worm-like intergrowths with quartz: very distinctive. At Shane's Hill, there is a rock called rhyodacite in which the feldspars occur as large, blocky, isolated crystals, very different to the feldspars at Slieve Gullion. And feldspar occurs as a constituent in the sedimentary Ordovician rocks of Armagh.

County Fossil: Shark's Tooth

Armagh was once full of sharks. Limestones laid down in the tropical shallow seas of the early Carboniferous Period around Armagh city are anomalously rich in shark teeth. Really quite remarkable. Other parts of the sharks we don't see because sharks are cartilaginous, whereas their teeth are hard calcium phosphate and, hence, capable of being preserved.

Cavan

The Prehnite County

Cavan was the first area in Ireland or Britain to have had demonstrated the existence of a regional scale of low-grade metamorphism known as the prehnite–pumpellyite grade (or 'facies' to use the technical term for 'grade'). This was big news in 1978 and Graham Oliver, who discovered this, made it into the prestigious science journal *Nature*. This type of metamorphism had been predicted for the Ordovician and Silurian rocks of the northern part of Ireland and for related parts of Britain but had never been found. When it was discovered in Cavan, it opened the doors to finding it across huge swathes of these islands and, thenceforth, all over the world.

County Rock: Glacial Till

Glacial till is the unconsolidated muds, sands and boulders that were dumped, channelled, mounded and ridged on the surface of Ireland as a result of glaciers either physically moving it or melting and depositing it. Till is a hugely important modern rock in Ireland both for agricultural and engineering purposes.

County Mineral: Baryte

The county rock of glacial till, frustratingly, tends to obscure the bedrock to a large extent, so finding a county mineral and fossil is not so easy. However, at Muff there has been reported a 4cm-wide vein of baryte (barium sulphate), and this is a fine county mineral to have.

County Fossil: Atavograptus atavus

Fossils are rare in Cavan (due to the till again). But some Silurian Period shales in the southern part of the county contain graptolites, one of the few fossils collectable in Cavan; a graptolite is a colonial

Paleozoic animal that floated around the ancient oceans, each individual living in a kind of tiny cup but all connected together by a chitinous support. I have chosen the graptolite *Atavograptus atavus* as Cavan's county fossil.

Derry

The Chalk County

Some wonderful exposures of chalk occur along the north coast roads of Derry. And in the chalk is a range of fossil animals not common in the rest of Ireland. Chalk also has the element of mystery to it: how and why did chalk form in such quantity during the Cretaceous Period?

County Rock: Chalk

Chalk is actually a remarkable rock. Ulster chalk is 99 per cent or more pure calcium carbonate, is made from countless trillions of fossil coccoliths that all lived and died during the Late Cretaceous Epoch, and is noticeably harder than its famous (softer) southern England brethren. It is likely that chalk once covered most of Ireland, but now it is almost all gone, except where preserved under the Antrim basalt, plus one enigmatic outcrop at Ballydeanlea, County Kerry.

County Mineral: Gyrolite

This mineral is a hydrated sodium–calcium aluminosilicate hydroxide and is usually found as part of the infilling of basalt vesicles (once gassy holes/cavities). At Sounding Hill quarry, just outside Magherafelt, there are gyrolites that occur as densely packed flakes, pearly white in colour, lining some of these 'holes' in the basalts.

County Fossil: Belemnite

Fossils in Derry are very restricted because most of the rocks in this county are too old or not of the right type. However, one characteristic fossil from the coastal Cretaceous Period chalks, beneath the basalts, is the bullet-shaped belemnite. This is the central hard part of an extinct cephalopod, like a squid. The rest of the animal, being soft tissue, tends not to survive fossilisation.

Donegal

The Granite County

The Donegal batholith, comprising its six different plutons, is world-famous (see Chapter 12). The general public may not realise just how geologically famous these plutons are. The granites of Donegal are regarded as classics, and feature very widely in all manner of books and articles on granite.

County Rock: Granite

As stated in the above nickname for Donegal, granite is the county's superstar rock ('rocks', really). Not only are there many different varieties, but the way they have been emplaced (i.e. arrived where we see them today by all the different, but possibly related, intrusion mechanisms) is legendary among geologists. They have a long history of geological attention, and they have some associated collectable minerals, such as beryl.

County Mineral: Kyanite

With a name derived from its often arresting colour (cyan-blue) and being a characteristic mineral of Donegal, kyanite, a structural variant (polymorph) of aluminium silicate, makes a great county mineral. It occurs in the rocks immediately surrounding both the Ardara Granite and the Main Donegal Granite. Large, two-fist-sized masses of bladed sky-blue crystals are known. Indeed, I have seen some wonderful kyanites in schists (metamorphosed sandy muds) used in the building stones of a castle in north Donegal. Geologists often consider kyanite to be a mineral indicative of high-pressure geological environments. However, some of the kyanites found in Donegal clearly grew into relatively shallow underground open spaces in vugs or gas pockets, which couldn't possibly be at high pressure. How?

County Fossil: Rhizocorallium

The tropical limestones and limey sandstones of the Carboniferous rocks that encircle Donegal Bay contain evidence that large sea worms once feasted in the originally soft muds. The traces of these worms are called *Rhizocorallium,* and they appear as fossilised burrows in the rocks.

Down

The Aquamarine County

Translucent, sea-green aquamarine beryls are one of the characteristic minerals that come from County Down, making it well known for over a century. Beauty and historical significance make the name 'Aquamarine County' a good choice for Down.

County Rock: Permian Sandstone

In Irish terms, this is an uncommon rock type and, therefore, quite distinctive. Most of Ireland's Permian Period rocks have been stripped away by the vagaries of history. But on the northern coast of Down are Permian desert sands (now sandstones) and other rocks associated with arid conditions, such as those produced from periodic flooding, scree deposits and lake evaporation. Ireland during the Permian was located where the Sahara Desert is now.

County Mineral: Aquamarine Beryl

Aquamarine is the translucent pale-blue variety of beryl (beryllium–aluminium silicate) and has been found for very many years in the Paleogene granites of the Mourne Mountains. Crystals can still be found today, but the biggest and best crystals were possibly all found in the nineteenth century. In fact, the nineteenth-century finds have been described as the best aquamarines ever found in Ireland or Britain (Tindle, 2008).

County Fossil: Chirotherium

This is the name given to the footprint trace of an extinct early reptile from the Triassic Period that was found in the Triassic sandstones of Scrabo quarry. The beast itself might have been a *Ticinosuchus*. A great county fossil to have.

Fermanagh

The Cave County

With Marble Arch Cave and an internationally recognised geopark in its borders (with Cavan), Fermanagh can only be called the Cave County.

County Rock: Stalactite

Fermanagh is like a Swiss cheese: full of holes. The most dominant rock type is Carboniferous limestone and, at the end of the last Ice Age some 10,000 years ago, melt-water dissolved out caves and underground tunnels from the limestone. These spaces in turn hosted stalactites (that hang from the roof) and stalagmites (that grow up from the floor) as, over millennia, drips of calcium carbonate and rich water precipitated their dissolved load in the form of phantasmagorical cave formations.

County Mineral: Dolomite

Dolomite is a common calcium–magnesium carbonate. However, Fermanagh hosts the most extensive dolomite deposits in Northern Ireland, so can be well pleased with that!

County Fossil: Delepinea destinezi

Specimens of this little brachiopod shellfish species from Kesh, County Fermanagh – named not after the Men of Destiny but after a diligent Belgian geological technician, Pierre Destinez (1847–1911), in 1915 – were featured heavily in a 1966 review of this sea creature's genus description and characteristics. Fermanagh is particularly blessed with brachiopods; at one locality over 80 different species have been recognised.

Monaghan

The Drumlin County

The word 'drumlin' is one of the few scientific technical terms to be derived from Irish, in this case *druim* (mound; *druimnín* – little ridge). Drumlins are formed through a combination of processes involving subglacial deposition of the comprising till and then some erosion of that. Drumlins tell of the last direction of ice flow, and Monaghan is a great county for drumlins.

County Rock: Permian Red Mudstone

During the Permian Period, Ireland was enjoying hot, arid conditions – not unlike the modern Sahara. But most of our rocks of this period have been eroded. One of the few good outcrops we have in the Republic of Ireland from the Permian (and Triassic) periods is in and around the geologically well-known Kingscourt gypsum mine at Knocknacran. Mudstone may not seem anything special, but when you actually see the blazing, dramatic reds of the mudstones that overlie, and stunningly contrast with, the white gypsum layers at Knocknacran gypsum opencast mine, it will knock your socks off.

County Mineral: Rhodochrosite

Rhodochrosite is a manganese carbonate that, in worldwide terms, is quite common. But in Ireland it is very rare. The only fully confirmed occurrence in Ireland of this lovely pink mineral is from Logwood Hill, Calliagh, where it occurs in a small iron–manganese deposit in a quarry there. As the only unambiguous occurrence in Ireland, rhodochrosite is a good choice for Monaghan's county mineral.

County Fossil: Lueckisporites virkkiae

Because Monaghan hosts rare, in Irish terms, examples of Permian Period rocks, I have chosen as the county fossil a late-Permian Period plant spore. *Lueckisporites virkkiae* is one of the most important spores that can be collected (albeit by special means) from the Kingscourt area. It is probably the spore of the late-Permian conifer *Ullmannia bronni,* which lived close to the (transient) warm Monaghan waters of the time, into which the hot Permian winds blew its spores. *L. virkkiae* has also played an important role in tracing (correlating) the layers of some of the Permian rocks right across Europe.

Tyrone

The Geologist's County

Tyrone is exceptional. It has an extraordinarily diverse range of rock types and rocks that span almost the entirety of Ireland's geological history, from the Precambrian right up to the Paleogene Period. It has coal and it has gold. It has bits of an old ocean floor, as well as rocks laid down in a desert. And it has enough complexity to keep the structural geologists very happy. Tyrone has pretty much everything.

County Rock: Pillow Lava

Tyrone contains the deformed remnants of the ancient Iapetus Ocean squashed into it. This ocean existed when Ireland was separated into two physically different halves; the closing of this ocean 400 million years ago, in the southern hemisphere, is what brought Ireland, as we now know it, together for the first time. On the floor of that ocean, lava was erupting, and when lava extrudes underwater it tends to form 'pillow' shapes – we can see such pillow lavas preserved at Craig. These lavas play a crucial role in an epic story.

County Mineral: Pyrite

Pyrite, aka 'fool's gold', is a very common variety of iron sulphide and is found in many counties. But in one case in Tyrone, the fool was no eejit. It was pyrite occurring in quartz veins in Dalradian-age (late Precambrian or a bit younger) metamorphosed sediments around Curraghinalt, near Gortin in the Sperrin Mountains, that helped lead Northern Irish geologist Garth Earls to find real gold.

For although pyrite can fool the inexpert eye, the expert eye knows that gold itself is often associated with pyrite. For fool's gold leading to real gold, pyrite is Tyrone's county mineral.

County Fossil: Calyptaulax brongniartii

Tyrone has the best fossiliferous Ordovician sandstones in Ireland. And one of the fossils to be found in these rocks, which occur directly to the east of Pomeroy, is the lovely shallow-water trilobite with the tricky name, *Calyptaulax brongniartii.* This trilobite had a pair of schizochroal eyes, i.e. each eye had many lenses, like that of a fly.

LEINSTER

Carlow

The Pegmatite County

The pegmatite vein rocks on the eastern margins of the Leinster Granite in southeast Carlow have been the cause of much prospecting by mineral companies, and a source of great academic debate about whether these coarse mineral veins originate from the granite, as a last-gasp expelled crystallisation residue, or from the surrounding metamorphic rocks, as the partially molten part of these rocks, the hot liquid being sucked into the space made by opening cracks. This debate currently rages. But both their size and the exotic suite of minerals they contain means the Carlow pegmatites have been recommended for national heritage protection.

County Rock: Pegmatite

There is a certain geological irony with Carlow: it has the highest proportion of granite in its borders of any county in Ireland, yet most of it cannot be seen and it's not at all famous. However, it does have some excellent coarse veins associated with granite known as pegmatites, and Carlow's are special. Pegmatites often contain very large crystals, and around Aclare and Myshal they contain large crystals of exotic minerals such as the lithium mineral spodumene, which can reach an eye-boggling 30cm or more in length. And there are also the rare minerals of columbite, tantalite, microlite, phosphosiderite, lepidolite …

County Mineral: Andalusite

Andalusite is a purple/pink elongate form of aluminium silicate polymorph (it has the same chemistry but a different structure to that of kyanite – see Donegal above) that will forever, at least in Ireland, be associated with Carlow. Why? Because Carlow has within its borders a lot of andalusite; and andalusite is used to make refractory materials, is involved in the production of iron and steel, and was the source of some hot and protracted local politics between 1989 and 1992. The andalusite deposits around Tomduff were the subject of a battle between Navan Resources plc, which wanted to mine (or test the possibility of mining) the andalusite around Tomduff and Mount Leinster, and the local anti-mining group, the Mount Leinster Mining Awareness Group, which was established to oppose the company's plans. After much argument and counter-argument, Navan Resources lost. But it kick-started the political career of Mary White, who went on to join the Green Party.

County Fossil: Lonsdalaeia floriformis floriformis

Carlow is mostly granite, with some metamorphic rocks and pegmatites. No fossils in those rocks. But in the west of the county is Carboniferous limestone, and at Bannagagole limestone quarry, near Oldleighlin, was found only the second occurrence in Ireland of the colonial coral (with three parts to its name), *Lonsdalaeia floriformis floriformis*.

Dublin

The Muscovite County

More than any other county, Dublin is dominated by the variety of mica known as muscovite, because muscovite is an integral part of all the local granites. And all the local granites are seen every day, sparkling and winking, in hundreds of buildings and walls. Really big flakes (between 2cm and 7cm), uncommon by international comparison, characterise some of the Dublin granites. The Dublin granites' muscovites were among the first in the world to be shown to possess growth-zone structures: the mineral equivalent of tree rings (Sollas, 1891). A century later, in 1991, my own investigations helped settle a long-running debate on whether muscovite in granite is part of the original magma or some sort of later overprint. I can tell you that they are magmatic. See Plate Figure 9, and the front cover of this book.

County Rock: Lambay Porphyry

'A rare auld Dubbelin rock', Lambay Island is the remains of a huge Ordovician Period volcano. The highly distinctive, and geologically famous, rock that comes from here is mid-dark green, has a fine-grained matrix but large, blocky, well-shaped crystals set through it, like rectangular, pale-green currents in a darker-green cake. Known as Lambay porphyry ('porphyry' being a type of extrusive volcanic andesite rock), it was first described in 1878. It featured in innumerable geology books in the late nineteenth and twentieth centuries, and was widely used in Neolithic stone axes.

County Mineral: Zoned Muscovite

Muscovite is a platy, sun-glinting member of the mica family. In Dublin, crystals in the local granite can be up to 7cm long, which is the current world record (personally witnessed in Walshe's quarry in Stepaside – I'll never forget that day). When observed with the naked eye the muscovites look good, but nothing exceptional. However, when placed between two pieces of polaroid plastic and viewed with a hand lens, or (best) under a polarising microscope, the visual colour feast from growth and corrosion features that these muscovites display is quite breathtaking (see Plate Figure 9). The Dublin-zoned muscovites have graced the cover of a magazine (*Geology Today*, 1992, Vol. 8, part 6) and have been written about in textbooks on minerals as well as in textbooks on granites. There are billions of these muscovites, and yours truly loves each and every one of them.

County Fossil: Orthoceras Sancti-Doulaghi

This animal was a good-sized cephalopod (related to the modern squid) that could be at least 24cm long. It enjoyed swimming in the warm waters of the Lower Carboniferous around St Doolagh's, now in Balgriffin, on the north side of Dublin city. This magnificent beast was discovered and named in 1897 from the exceptionally fossil-rich limestone quarries that once operated in the vicinity of St Doolagh's Church. Many fossils from the St Doolagh quarries made it into publications and museums. St Doolagh himself was a north-Dublin, Malahide-based, twelfth-century anchorite monk. And there are not many fossils named after Northside Dublin monks.

Kildare

The Volcano County

There are more old volcanoes in Ireland than one realises. But only very occasionally do they still look anything like a volcano. Kildare has a nationally well-known volcano in the Hill of Allen (actually Grange Hill), which dates from the Ordovician Period and is made of extrusive andesite lava and ash-fall tuffs (related in this way to Lambay Island, Dublin). The county fossil (see below) occurs in rocks just a smidge younger.

County Rock: Caliche

Caliche is the name given to a hardened, cemented layer of soils, sands or gravels, and forms at the surface in arid or semi-arid climates. Some 360 million years ago, Kildare was an arid desert, and rocks from this time can be found in a crude north–south semicircle around the east side of the Chair of Kildare (Carrickanearla). And in these Upper Devonian desert rocks can be found caliche nodules.

County Mineral: Marcasite

Marcasite is a polymorph of iron sulphide (as is iron pyrite) and is not a rare mineral. However, at the Harberton Bridge iron–lead–zinc deposit, marcasite occurs in quite a distinctive form not reported, to my knowledge, anywhere else in Ireland. Tindle (2008) describes marcasite from Harberton Bridge as occurring as 'a mass of lamellar hedgehog-like crystals'. Distinctive enough to be a county mineral.

County Fossil: Atractopyge verrucosa

Kildare's county fossil is an Ordovician trilobite that can be found in the limestones of this age at the Chair of Kildare (Carrickanearla) – a very useful fossil in determining the age within the Ordovician. Those with an interest in Latin might like the name, which translates as 'warty arrow-shaped buttocks'. So, if you see such buttocks in the rocks, twerking to catch your attention, grab them and bring them home.

Kilkenny

The Tree Fern County

Kilkenny is exceptionally rich in rare and important fossils. The county fossil is a stunning ancient amphibian; there is also an arthropod *Lagerstätten* (see Chapter 4). But Kilkenny is also known the world over (geologically, anyway) for its really ancient Devonian plants, which were found in a small quarry in Kiltorcan, near Ballyhale. These plants included giant tree ferns of the like we just don't have today, such as the Irish-named *Archaeopteris hibernica*. But plenty of superbly preserved plants and animals came out of the Kiltorcan quarry and many are yet to be properly described.

County Rock: Anthracite Coal

High-quality anthracite coal really only occurs in one place in Ireland and that is in the coal seams of the Castlecomer Plateau. Coal mining in Kilkenny goes back several centuries and only came to an industrial end with the closure of the Deer Park Colliery in 1979. However, small (cottage) operations still exist.

County Mineral: Tourmaline

Tourmaline from the banks of the River Nore at Inistioge has been known since at least 1837. Tourmaline occurs all down the margins of the Leinster Granite, and this is one of its most southerly outcrops. But this tourmaline does not occur in veins, which would be the norm – it occurs as layered beds, as if it were a sedimentary rock. And, in essence, it is. This layered tourmaline is the result of boron (the key element in tourmaline) having adhered to clays that were deposited on the old sea floor of Iapetus, this mix being subsequently metamorphosed into tourmaline later when that ocean closed and the two continents that came together to form Ireland collided.

County Fossil: Urocordylus wandesfordii

A rare and internationally important collection of eight ancient fossil amphibia genera was discovered in 1866 in the Upper Carboniferous Jarrow Coal Seam in the Kilkenny coal fields on the Castlecomer Plateau. These made the news because they were, at the time, the oldest amphibians known. One of the biggest of these was a creature, almost literally from a former black lagoon, called *Urocordylus wandesfordii* (see Plate Figure 10).

Laois

The Erratic County

An erratic is a large lump of rock that is foreign to the area in which it is found. Leaving aside Celtic giants as a causative agent, erratics tend to have been moved by ice, probably at the end of the last Ice Age, some 10,000 years ago. And Laois is a really great place to go hunting Galway granite! Dozens of huge boulders of Galway granite lie scattered all across the Laois countryside, having been brought this far east by the vast ice sheets that once plucked, scoured and left unconsolidated deposits all over Ireland. Go and have a picnic at the great granite erratic perched 300 metres up at the southern end of Gorteenameale, for example.

County Rock: Ironstone

The Upper Carboniferous ironstone of Laois is a very impressive rock. There may not have been much of it, but what was there was exceptional in its ore grade: three tons of this ironstone could yield one ton of iron, which is a phenomenal return. If only we had had more of it for industrial purposes! This iron ore tended to be high in sulphur. Additionally, the waters that percolated through some of the ore beds were used at one stage for curative purposes. Ah, you can't beat drinking and bathing in good old sulphurous iron waters!

County Mineral: Hematite

With ironstone being such an important county rock, we'd better include the main mineral that goes to form it: hematite (iron oxide). This iron mineral, with or without some other associated iron minerals, gives ironstone its weight, its colours (browns and oranges) and its value.

County Fossil: Actinocrinus

The graceful Carboniferous animal that was the sea lily *Actinocrinus* is a fitting county fossil, as it is found in some abundance in the Ballysteen Limestone mined at Knockacoller quarry, near Castletown. In the past, very fine heads of these sea lilies were found here. The limestone rocks of this quarry and its immediate surrounds preserve to an excellent degree many types of ancient sea life and are well worth a visit for any budding 'fossil head'.

Longford

The Inlier County

An inlier is an area of older rock surrounded by younger rock, and Longford gives its name to the most famous one in Ireland, known internationally as the Longford–Down inlier. This inlier is a big one and actually extends well beyond Longford, stretching across to the far edges of Down and even extending into Scotland. This inlier is a group of Ordovician and Silurian seabed sediments that were squashed and faulted during the closing of the Iapetus Ocean

400 million years ago, and whose history is vital for understanding that ocean's closure.

County Rock: Greywacke

The greywackes of Longford tend to be ... green. Greywackes (the name derives from the German *Grauwacke*) are layers of muddy sands that were deposited as a result of underwater sediment avalanches careering off down the continental slopes of the day (those of Laurentia) into deeper water, the sediment settling out in a characteristic way as it does so. The greywackes of Longford tell an important part of how the two halves of Ireland came together some 400 million years ago because these greywackes are the ones that got caught up in the continental collision.

County Mineral: Cerussite

Cerussite is a lead carbonate mineral, usually white to cream in colour, and in Longford it is the most dominant of the so-called secondary minerals (i.e. those that form as a result of alteration of the original minerals in a mineral deposit) in the lead–zinc deposit at Keel, Ardagh. The cerussite, being a lead-based mineral, forms on the edges of the primary lead mineral, galena.

County Fossil: Ortonella

Ortonella is a relatively unusual fossil that is found in the Lower Carboniferous rocks of Longford. It is the name given to crudely spherical clumps of ancient blue-green algae (oncoids of cyanophytes, for you lovers of words) and has been found in Ardnacassagh quarry. Limestone from this quarry was extensively used in constructing the Longford bypass. Some of you probably drive over *Ortonella*s every day.

Louth

The Suture County

Louth is extremely well known in geological circles the world over as the site of a giant stitch-up. And the stitching in question is that of the joining together some 400 million years ago of two continental masses (Laurentia and Avalonia) via the disappearance of the intervening ocean – the Iapetus Ocean. The join where the two continents meet, which extends for several thousand kilometres from Norway to the southern Appalachians in the US, is known as the Iapetus Suture. It runs right through Louth.

County Rock: Lamprophyre

An excellent example of a lamprophyre dyke, several metres across, outcrops at Clogher Head. A lamprophyre dyke is an intrusion of an unusual type of igneous rock, very rich in potassium but also iron and magnesium, and usually containing large crystals of dark mica. Lamprophyres are thought to form as a result of odd melting processes operating when ocean floor is subducted beneath continental crust. Which is exactly what happened here 400 million years ago (see Louth nickname above). Louth plays host to not one but three different types of lamprophyre rock: minette, kersantite and spessartite.

County Mineral: Neptunite

Any county with a mineral that is unique in Ireland or Britain is lucky indeed. Louth is one of those counties. Neptunite is a very rare titanium silicate mineral that also contains potassium, lithium, manganese and other elements. It is a deep red colour and occurs as small flakes between other mineral grains in a vein of igneous syenite rock that cuts through some local limestones near Barnavave, Carlingford. The neptunite probably crystallised as a result of some of the limestone being digested in the syenite magma to produce a very unusual composition of melt that crystallised a very unusual type of mineral assemblage.

County Fossil: Cyclopyge bumasti

This extinct animal is a type of trilobite, one of those woodlouse-looking marine arthropods that were so abundant during the Paleozoic Era. An excellent county fossil: it is inherently rare. It seems to be native only to County Louth, where it is found in rocks in the Oriel Brook, near Collon.

Meath

The Galena County

Meath is host to Europe's largest lead and zinc mine, located just outside Navan town. Galena, which also contains small quantities of silver (almost all Irish galenas contain some silver), is the main lead mineral found here. The mine is very important to Ireland's national economy and is, because of its size and quality, famous in the geological world and in international mining circles.

County Rock: Syenite

Syenite is an uncommon igneous rock, rather like granite but without the quartz. It is mainly, and most often, a mix of potassium feldspar, pyroxene and hornblende. Syenites are not common in Ireland, yet Meath boasts at least three occurrences – all slightly north or northwest of Navan and intruding into Ordovician rocks. Furthermore, anyone who wants to collect this odd rock for their mantelpiece can find in Meath at least three varieties spread across the three occurrences. Quite special.

County Mineral: Argyrodite

This is an extremely rare mineral: it is a silver–germanium sulphide. It is so rare in Ireland that only one single, solitary tiny grain of it has ever been reported, and that was from the Navan lead–zinc ore body. When you have something unique, make sure people know about it.

County Fossil: Dicranograptus clingani

This animal is a species of Ordovician graptolite and was chosen because graptolites are the main source of age dating and of correlating (and thereby making sense of) the pre-Carboniferous rocks in Meath. And it looks different enough from Cavan's graptolite.

Offaly

The Tufa County

Probably no county in Ireland has a better display of natural tufa than Offaly. Tufa is 'young limestone' formed during the Holocene Epoch (last 10,000 years) and on land (not in the sea) by calcite precipitation from carbon dioxide-rich rivers, small waterfalls or from water percolating out from any limestone-rich rock or glacial till. A most impressive tufa, forming a series of cascades, is to be found in the Millpark brook near Mount Joseph Abbey, while one of the largest tufa mounds (12 metres x 8 metres) is to be found by the Aghagurty River.

County Rock: Clonony 'Marble'

Clonony, a small village in west Offaly, features a castle once owned by Thomas Boleyn, whom King Henry VIII once had in mind as a prospective father-in-law. Clonony was well known in the nineteenth century for a quarry that produced a very decorative variegated orange (with purple and grey patches) 'marble'. It is not a true metamorphic marble but a coloured limestone that takes a good polish. Clonony 'marble' has been used to very dramatic effect in Trinity College Dublin's Museum Building, for example, where it forms some of the huge internal pillars there. Definitely one of the most nationally famous rock types to come out of Offaly.

County Mineral: Calcite

A very large percentage of Offaly is underlain by Carboniferous limestone. Limestone is a rock made of the mineral calcite (calcium carbonate), and when limestone gets fractured or deformed – as has happened in Offaly – then the dirty, impure calcite of the limestone rock is dissolved and reprecipitated as pure white to clear mineral calcite in veins and pockets, and often as very good crystals in many different shapes (calcite is mineralogically notorious for its many different crystal shapes).

County Fossil: Phillipsia

A happy yet sad story for this fossil. Happy, because the trilobite species of the *Phillipsia* genus can be found in the Lower Carboniferous limestones of Offaly, and is a handsome little chap (or chapess) about 2.5cm when fully grown and well worth seeking out in a limestone near you. Sad, because this trilobite has the distinction of being the 'Last of the Trilobites': with *Phillipsia,* a 300-million-year-old dynasty ended. But that wasn't to happen until the end of the Permian Period. And Offaly was not responsible. The *Phillipsia*s during the Carboniferous Period were all happily truffling about on Offaly's then-tropical sea bottom.

Westmeath

The Xenolith County

Westmeath is literally almost all limestone (see county rock below). But there are some exotic strangers – the xenoliths – that have made their home here and are what really make Westmeath special. For example, at Clare Castle and in the bedrock of the Dungolman River, you will find rocks that have been shot up through a (long extinct) Carboniferous volcanic vent; they are visitors from the deep, lower crust some 30 or 40km beneath Westmeath. These rocks, termed xenoliths, are fragments of normally deeply buried, well-hidden, considerably metamorphosed, and very ancient, Westmeath. Xenolith just means 'strange rock'. The very names of these 'strange rocks' sound exotic: banded pelitic paragneisses, khondalites, kinzingites, plagioclase–hypersthene basic xenoliths. They are all very important because they give us a glimpse of deepest Ireland, unseen and unseeable by any other means.

County Rock: Limestone

In general, Carboniferous limestone is one of Ireland's archetypical rocks and no county has proportionately more of it than Westmeath. The only trouble is that most of it is buried beneath either bog or glacial till.

County Mineral: Millerite

An unusual occurrence of the bright, brassy, needle-like nickel suphide mineral millerite was found at Ballinalack, some 15km northwest of Mullingar, in the sub-economic lead–zinc deposit there.

County Fossil: Siphonodendron

The county with the most Lower Carboniferous limestone is assigned one of the most characteristic Lower Carboniferous Irish fossils: the colonial coral *Siphonodendron*. These are widespread in the limestones that formed in the shallower waters of the time and are a mass of slightly separated branching tubes that can seem to be gently radiating or in parallel to one another – they look a little like fossilised macaroni.

Wexford

The Fossil County

The exceptional fossils seen in the limestones at the end of Hook Head are nationally important and are one of the few such sites actually protected by law. Do not collect them. But do go and see them – they are one of our national outdoor treasures.

County Rock: Amphibolite

Not just any amphibolite! Amphibolites usually form as a result of the metamorphism of iron-rich igneous rocks: basalts, for example. And in south Wexford we can see amphibolites around Rosslare Harbour and near Greenore Point. These rocks are part of what geologists term the Rosslare Complex, and they are special. They are the oldest rocks in the south of Ireland (some 620 million years old) and are part of the old, southern continent of Avalonia, which crashed into the northern continent of Laurentia 400 million years ago, causing the two halves of Ireland to come together for the first time. If you stand on rocks south of a line from Rosslare to Kilmore Quay you are, in a real sense, on a completely different continent to comparable rocks north of a line from Drogheda to Clare (such as in Galway or Donegal). Amazing.

County Mineral: Chromite

Chromite is not common in Ireland, and Wexford probably has the best of it. Chromite is iron–chromium oxide, and in Wexford it occurs as disseminated grains and as thin beds as part of a block of serpentinite rock at Cummer. This serpentinite block can be shown to have been shoved onto Wexford by tectonic forces that operated during the closing of the Iapetus Ocean, and the subsequent crashing (in slow motion) of continents some 400 million years ago. Cummer is getting better known nationally as a potential source of both chromium and nickel.

County Fossil: Ediacaria booleyi

Ediacaria booleyi is my most controversial choice for this chapter because geologists are currently debating whether this is even a real fossil, albeit a trace fossil. *If* it is, then it is extraordinary and of world significance. In the shales at Booley Bay, south Wexford, occur mid- to late-Cambrian dome-shaped structures up to 18cm wide, concentrically zoned, and that seem to show tubules. These are thought (by some) to be related to the soft-bodied Precambrian-age Ediacara fauna originally described from Australia. The Wexford specimens potentially represent the youngest such occurrence of this Ediacara fauna, which would have international significance. Even as an object of scientific controversy (whether Ediacaran or not), *Ediacaria booleyi* is a worthy county fossil.

Wicklow

The Gold County

Gold occurs in a number of Irish counties, but Wicklow trumps them all. It was in Wicklow that we had one of Europe's rare, and possibly first, gold rushes (in 1795). This produced Europe's largest ever gold nugget (estimates vary from 624g to 750g, but the lower figure seems most likely) and a total estimated weight of recovered gold of some 300kg. Oh yes! Wicklow is the Gold County.

County Rock: Coticule

This odd-sounding name comes from a rock originally described in Belgium and that was used for sharpening shaving razors (whetstone). But Wicklow has even better coticule than the Belgians. Coticule is a pinkish rock made of tiny spessartine garnets and quartz, and it normally occurs as a series of bizarrely folded layers. Outcrops can be seen at Lough Tay and on the sides of Glendalough, among many other places. It is an exceptional rock type: it was produced by the metamorphism of rocks derived from ocean-floor black smoker emanations; it is laterally extensive for hundreds of kilometres, through the east side of the Leinster Granite up into Scandinavia in one direction, and down the Appalachian Mountains (United States) in the other. It is also a predictor of all manner of metal deposits, including gold, lithium, tungsten, copper, lead and zinc. It is the most extraordinary rock that no one has ever heard of.

County Mineral: Gold

Wicklow was given gold as its county mineral, and rightly so. The gold occurs mostly in streams as placer deposits, but probably got there through weathering of gold-rich chalcopyrite and local ironstones, with a smaller contribution weathering out of quartz veins. Did the ancients know of this gold and did they use it to fabricate our wonderful Early Bronze Age gold artefacts? See Chapter 9 to find out.

County Fossil: Oldhamia

Wicklow is rich in geological treasure and hosts Ireland's oldest fossils. These are not actual body parts, but the traces left by soft-bodied animals, probably worms. *Oldhamia* are centimetre-sized sets of tiny, often radiating, burrows (e.g. *Oldhamia radiata*). They were formed during the Cambrian Period, roughly 530 million years ago, on the bottom of a then-quiet, muddy Iapetus Ocean floor. They have been collected from Bray Head (Wicklow) and Howth Head (Dublin). These probable worm traces were named in honour of nineteenth-century Dubliner and Geological Survey of Ireland geologist Thomas Oldham.

MUNSTER

Clare

The Limestone County

Overground or underground, Clare limestones are impressive. Whether it be the gently warped limestone benches of the Burren; the mighty Great Stalactite of Pol-an-Ionain, one of the biggest free-hanging stalactites in the world and now part of a tourist walk-in show cave (not like the old days when one had to crawl and squeeze through dark, underground, water-filled passageways for hours getting bruised from head to toe, which I will never forget doing); the Poulnagollum–Slieve Elva cave system, which, at more than 14km, is the longest cave system in Ireland; or the many text-book karstic features on show. It is the range, the diversity and the quality of Clare's limestones that makes them first-class.

County Rock: Liscannor Flagstone

Probably the most distinctive rock to come from Clare is the well-known Upper Carboniferous Liscannor Flagstone (from Liscannor), characterised by thick, rambling, sinuous ridge patterns. The rock is a muddy, mica-rich sandstone, formed originally as a mouth bar in a river delta, and these patterns are trace fossils called *Psammichnites* (formerly called *Olivellites*). The origin of the *Psammichnites* is as a type of feeding pattern from either a large worm or a mollusc or a burrowing arthropod. One suspects it might be a worm, which would not fossilise in this type of rock. In any case, the beastie responsible has never been seen.

County Mineral: Fluorite

Purple and clear crystals of fluorite (calcium fluoride) occur at Sheshodonnell East lead mine, and its waste dumps, in the Burren. I have chosen fluorite as the county mineral for Clare not just because some lovely material comes from here but because of a happy personal memory. It was here, on an educational school trip to the Burren, that I collected my first Irish fluorites ... while I should have been writing notes about sheep farming or something.

But I was otherwise distracted, and fluorite has always been one of my favourite minerals. Fluorite occurs at other locations in the county, too: Ballyganner South, Castletown, Gleann na Manach and Skancullin, to name but a few.

County Fossil: Reticuloceras stubblefieldi

County Clare is one of the best counties to find Carboniferous goniatites, the coiled-shell relatives of the squid. Good for goniatite collecting, and also good as a geological marker to help geologists map out the rocks of the Cliffs of Moher, is a layer of black marine shale with a limestone bed that contains within it plentiful examples of this goniatite. But do be careful if you go hunting the *Reticuloceras*.

Cork

The Old Red Sandstone County

An entirely appropriate nickname because no other county in Ireland has more Devonian Period (415–360 million years ago) desert sands and related desert rocks than Cork. And, most conveniently, they can be seen as the hills and mountains of the county because the original rocks were later buckled upwards by what is known as the Variscan orogeny, i.e. a mountain-building event caused, in this case, by Iberia and North Africa bumping into us at the end of the Carboniferous Period.

County Rock: Cork Red Limestone ('Marble')

More commonly called 'Cork Red Marble' (although it isn't a true marble), this rock is a pink to red limestone, sometimes displaying brecciation (broken-up) features, that is set in a deep-red mudstone matrix. It has been used for many years as a decorative building stone, and one only has to walk into the Museum Building in Trinity College Dublin to see the architectural drama created by pillars of 'Cork Red' when cut and polished. Still worked today, it is probably best seen at a quarry in Baneshane, near Midleton.

County Mineral: Corkite

This was a very straightforward choice as county mineral: corkite is the only internationally recognised valid mineral species to be named after an Irish county. No more reason is needed than that. Corkite is a rare lead–iron sulphate phosphate hydroxide that occurs either as brilliant translucent brown crystals or as minute yellow rhombs. The type area is at Glandore iron mine, although it has also been found at nearby Aughatubrid Beg mine.

County Fossil: Ostracod

Cork is the largest county in Ireland. And ostracods are some of its smallest fossils (usually about 1mm long or less), being found, for example, in the dark-grey mudstones, containing lenses of limestone within it, of the Old Head of Kinsale in West Cork. Although by no means exclusive to Cork, it is somewhat appropriate that the biggest county has the animal with the largest 'langer' in proportion to its body size in the *entire* animal kingdom, ever. A modern ostracod's sperm is ten times its own body length, and the thing that discharges it towards the female takes up one-third of its body volume. Ancient Corkonians may have been small, but they were mightily endowed.

Kerry

The Tetrapod County

Kerry became geologically world-famous for having, on Valentia Island, the second-oldest tetrapod ('four-footed') trackway in the world (just pipped by a site in Poland). And this is highly significant because it is evidence of the first four-legged animals (amphibians) ever to come out of the water and make tentative forays onto land. The Irish State has now purchased the site: it is a National Heritage Site and is protected in law. But visitors are very welcome. A national and international treasure.

County Rock: Inch Conglomerate

One of the most visually dramatic and geologically fascinating conglomerates in all Ireland can be viewed west of Inch spit on the Dingle Peninsula. A conglomerate rock is one made up of large rounded lumps of other (older) rock all cemented together: the old name was 'puddingstone'. The Inch conglomerate is a generally purple conglomerate, of Devonian age, containing rounded clasts of all sizes, from sand up to boulder, and of many different rock types, but mainly metamorphic. Geologists think that this conglomerate derived from the erosion of an old mountainous area that existed during the Devonian here and that once lay to the south of Dingle, i.e. there were once mountains where now there is sea.

County Mineral: Rock Crystal (Kerry Diamonds)

Kerry is famous for a great many things – ask any Kerryman – and one of those is the so-called Kerry Diamonds. These crystals are not, lest anyone be in any doubt, real diamonds, but they are superb, water-clear, beautifully formed crystals of quartz. Sometimes they can be a jaw-dropping 15cm in length and may even be doubly terminated, i.e. have crystal faces at both ends of the prism length. Locally, they can be prolific. The origin story of these very collectable quartzes is that they crystallised from very pure silica-rich fluids that had come from dissolving some of the Devonian sandstones, the fluid then precipitating the 'diamonds' in gashes and cracks in the sandstone rocks that had opened up because of regional-scale deformation of these sandstones. This was as a result of the 300-million-year-old Variscan orogeny, when Iberia, which had been separate, crashed into the rest of Europe, including us. Thank you, Spain, for the Kerry Diamonds.

County Fossil: Valentia Tetapod Trackway

A trace fossil of world importance – some 150 footprints left in originally damp sands by a 1-metre-long salamander-like animal some 385 million years ago (during the Devonian Period). This is evidence of evolution allowing formerly water-dwelling vertebrates to start inhabiting land for the very first time. Some of the first steps taken on land by any quadruped happened in ancient Kerry.

Limerick

The Mud Mound County

Limerick has a world record. It contains, in its western parts, the thickest sequence of Waulsortian limestones in the world, estimated to be about 1km thick. Waulsortian limestones are a special type of limestone – all of Lower Carboniferous age – and they occur in Ireland, Britain and continental Europe. They comprise large, topographic mounds of limey mud, which grew up from the deeper parts (below 200 metres) of the tropical sea floor of the time, and which also hosted rich faunal communities. These lime mud mounds are thought to have been produced and held together either by microbial mats or by mucus. So, one way of looking at this is that west Limerick has the biggest pile of fossil mucus on the planet.

County Rock: Ankaramite

This rock is a variety of volcanic basalt, but quite a special one: it contains a relatively high percentage of large crystals of pyroxene and a lower percentage of olivine compared to most basalts. In Limerick, one can find ankaramite lava in the area around Pallasgreen. These lavas erupted underwater and subsequently on land (land made from the accumulated previous lavas), and their particular and peculiar chemical characteristics are a sign that the continental crust at the time was not totally ripped apart but more gently torn. Nevertheless, Pallasgreen 335 million years ago was a dangerous place to stand.

County Mineral: Cacoxenite

Gorgeous things can come out of Limerick, and cacoxenite, a rare and (very) hydrated aluminium–iron phosphate hydroxide, is one of them. The only occurrence of this mineral in Ireland is very close to the old fort of Lismeenagh, Ballycormick, near Shanagolden. And, appropriately for its location at Shanagolden, the colour of cacoxenite is a golden yellow, appearing like mini crystalline sunbursts of radiating fibres. Spectacular!

County Fossil: Megalocerus giganteus

The iconic Irish mega fossil *Megalocerus giganteus* is better known as the giant Irish deer. A magnificent complete male skeleton, with its absolutely huge antlers intact, was extracted from the bog at Lough Gur. This specimen causes gasps from visitors entering the Museum Building in Trinity College Dublin (home of the Geology Department) where it is now permanently on display (see Plate Figure 11). Go and see it. Other complete specimens can be marvelled at in the National Museum of Ireland – Natural History on Merrion Street, Dublin (aka The Dead Zoo). This great beast roamed Ireland during the warmer intervals of the last Ice Age, along with bears (see Leitrim), mammoths and hippopotamuses.

Tipperary

The Silver County

Tipperary can boast the oldest documented mining in Ireland (i.e. proven from written records, as opposed to archaeology alone). Mining in the Silvermines area goes back to at least 1289, although it seems likely that it goes back to the ninth century and probably even before that. The earliest mines did seem to produce silver, but in more modern times the essential elements are those of lead and zinc. The silver seems to have been won from the common lead mineral galena (it being argentiferous). Remarkably, despite many centuries of mining, Tipperary remains an essentially rural county.

County Rock: Wenlock Siltstone

At first glance, this rock may not seem like anything special. You could not be more wrong. The Wenlock Epoch of the Silurian Period was from 428 to 423 million years ago and the siltstones in question are shallow marine siltstones, part of the Hollyford Formation, found in the hills south of Moneygall. But these siltstones are special. They host fossils of the world's oldest land plants (see '*Cooksonia*' below): the very first time self-supporting plants of any kind ever appeared on land anywhere in the world. And it was in Tipperary.

County Mineral: Gortdrumite

One of the few examples of a mineral species named after an Irish locality: for this reason alone it is a worthy county mineral. Gortdrumite is a rare copper–iron mercury sulphide, and the only place in Ireland or Britain where it is found is in Gortdrum mine, near Tipperary town. Crystals are usually less than 2mm long, a kind of blackish-red colour, and were formed at the relatively low temperature of around 150 degrees Celsius.

County Fossil: Cooksonia

Tipperary has something of unquestioned world importance in the evolution of life: *Cooksonia* (see Plate Figure 12). The very first plant ever to have grown on land. And that land was some 424 million years in the past as near-shore swamps or tidal flats in Moneygall (Slieve Phelim/Devil's Bit) during the Silurian Period. These first plants were at most 5cm high (usually much less), had a supporting vascular system, respiratory pores and a pair of sporangia waving at its top (looking exactly like tiny disco-ball head boppers). Tipperary hosted a momentous event in the evolution of life on Earth.

Waterford

The Copper County

Giving Waterford this nickname was not one of the more difficult choices, as copper mining has gone on here for centuries. The county contains a number of fascinating and sometimes rare copper minerals and it has the magnificent azure-blue copper mineral 'waterfall' in the mine under Tankardstown (see Chapter 13 and Plate Figure 2). In addition, Waterford has the whole Copper Coast Geopark to enjoy.

County Rock: Rhyolite Tuff

West Waterford was a very volcanic place back in the Ordovician Period. One of the dominant volcanic rocks was rhyolite, a silica-rich lava that either plugs up volcanic vents or explodes violently to produce rocks made of its own shattered products. At Dunhill Castle Quarry we see, after a short tramp through a bog and a farmer's slurry-pit overflow, a series of rhyolitic tuffs. These tuffs are deposits from volcanic ash clouds and include exploded-out lumps of solid rhyolite, but also rhyolite pumice. At Dunhill, these tuffs, which were originally airborne, show evidence of having been deposited underwater. So, there was sea next to the volcano. Think either the Philippines or Montserrat today, and that was Waterford during the Ordovician Period.

County Mineral: Lavendulan

This is a rare, often sky-blue, copper mineral (a hydrated sodium–calcium–copper arsenate chloride), and it is found in the Seven Dial Lode mine in Knockmahon as very small, pale turquoise-blue spherules (associated with other copper minerals) and on copper-mine waste dumps near Bunmahon. For Waterford, it was obvious that a copper mineral had to be chosen – and there were several candidates – but I picked lavendulan because, apart from it being both rare and appropriate, I liked the sound of the name. It sounds like a good county mineral.

County Fossil: Eirelithus thersites

It is not often one finds a fossil named after a country, but in Newtown Glen, in the Ordovician Period rocks of the Tramore Limestone, is a trilobite named after Ireland itself: *Eirelithus thersites*. This trilobite belongs to the Trinucleid family, looks vaguely like a woodlouse, and has been described as quite fast-swimming and eating whatever fell to, or lived on, the Ordovician ocean floor. Additionally, it has played a role in correlating these old limestones with equivalent rocks elsewhere and, hence, helping to unravel the detailed sequence and structure of rocks of this age.

CONNACHT

Galway

The Marble County

The seductive, swirly greens of Connemara marble are famous, and this marble is one of the very few genuine green marbles in the world. It is used for local and international jewellery and as a local and exported decorative building stone. More than enough reason for Galway to be the Marble County.

County Rock: Connemara Marble

Connemara marble is a true metamorphic marble. It is thought to have originated some 700 million years ago as a type of impure limestone containing iron and magnesium minerals; the original limestone was deposited on the shallow-water margins of the old Laurentia continent (see Chapter 1). Upon being metamorphosed, as a result of the collision of the Avalonia continent with Laurentia, these iron- and magnesium-rich limestones got transmogrified into the green (serpentine) marble we love today.

County Mineral: Molybdenite

This is a relatively unusual, soft metallic silvery-grey mineral (molybdenum sulphide) that has become well known since the 1980s as a result of a much-studied occurrence at Mace Head, near Roundstone. The molybdenite occurs in quartz veins and granite breccias that cut the Galway granite here.

County Fossil: Phanerotinus cristatus

Phanerotinus is an exceptionally large, rare and very distinctive gastropod that is found in the Carboniferous limestones of Galway. At 28.2cm diameter (measured from a more complete specimen in neighbouring Roscommon: shhh!), this is the world's largest Palezoic Era gastropod. Instead of having a tight spiral like that of a snail, it has a very open spiral like a flattened 2D spiral staircase.

And all along the outward-facing surfaces of its spiral are large 'dragon-like' curved spines. With its huge size and fearsome look, this was one marine snail you wouldn't mess with. No wonder it is the emblem of the Galway Geological Association.

Leitrim

The Bear County

The discovery of several fossil bears in a remote cave in Glenade made national and international news in the late 1990s. Apart from it being irrefutable evidence that bears roamed Leitrim in the geologically very recent past, the Glenade find was also one of the best that documented part of Ireland's lost megafauna from just after the last Ice Age.

County Rock: Iron Nodules

Leitrim once had the most extensive iron-smelting industry in Ireland because Leitrim holds (or held) Ireland's richest bedrock iron deposits. Slieve Aneirin is *Sliabh an Iarainn* ('iron mountain'). The iron is in the form of nodules, made largely of iron carbonate, that occur within the Upper Carboniferous shales in the hills of the Drumshanbo area.

County Mineral: Celestine

A celestial choice of mineral for Leitrim. Celestine (strontium sulphate) is not a rare mineral, worldwide, but is very uncommon in Ireland. Probably the best Irish celestines come from Carraun, near Kiltyclogher, where they have a slightly unusual mode of formation: the celestines occur as a mineral that has replaced a pre-existing mineral (in this case, well-formed gypsum) and can be up to 3cm long. Geologically, they are found in an evaporite type of rock (i.e. a rock formed through solar evaporation of a lake or a small sea, leaving behind deposits of various salts) associated with limestones of the Visean Stage of the Lower Carboniferous.

County Fossil: Bear

Initially discovered by cavers in 1997, bear fossils, including the remains of adults and cubs, were bravely collected for the nation from a cave some 200 metres almost vertically above the floor of the Glenade valley. This required the collectors to do a roped traverse, so as not to fall to their deaths. (How did the bears manage it?) Dating of the remains showed that the bears were living here as recently as about 4,000 years ago, which is contemporaneous with man in Ireland.

Mayo

The Amethyst County

To me, Mayo as the Amethyst County makes sense for aesthetic and historical reasons. Mayo is the premier county in Ireland for amethyst collecting, something that has been going on for some 180 years (and counting). Good specimens of the mineral itself are worthy to be on anyone's mantelpiece.

County Rock: Gneiss

On the Belmullet Peninsula are the most beautifully exposed ancient rocks on the Irish mainland – the Annagh Gneiss (pronounced 'nice'). This rock is some 1,753 million years old and is the geologically mangled remains of a series of really (really) old igneous rocks that have suffered ('enjoyed', we geologists prefer to say) several major periods of metamorphism. Only the gneisses on the island of Inishtrahull are older, but then only by some 30 million years.

County Mineral: Amethyst

The amethyst locality at the far end of Achill Island, at Keem Bay, is arguably Ireland's most well-known mineral deposit for collectors. First written about in 1832, it has been 'in the geological news' ever since, and very large crystals of this purple variety of quartz, ten inches long, were collected during the nineteenth century. Amethyst can still be collected today, with a bit of luck, but not of that size. And some enterprising locals sell it to visitors. One of the very few 'classic' mineral localities in Ireland.

County Fossil: Clorinda

Mayo is a geologically diverse county, but aside from its Carboniferous rocks, it is remarkably poor in fossils, with one notable exception. The Telychian Stage of the Llandovery Epoch of the (lower) Silurian Period that occurs in the Killary Harbour area of Joyce Country hosts outcrops of deepish marine sandstones that contain abundant examples of the brachiopod *Clorinda*. And apart from having a name one could give to one's daughter, this unusual brachiopod also has five sides.

Roscommon

The Pipe Clay County

Roscommon was once a major pipe-producing county in Ireland, supplying both local and national demand for inexpensive clay pipes on which one could happily puff away. The industry at Knockcroghery employed many dozens of people and lasted from about 1700 to 1880; after local clays were more or less exhausted, it robustly continued by using imported clays from England. Sadly, the whole pipe-making industry stopped abruptly in 1921 because Knockcroghery village was burned to the ground by the Black and Tans.

County Rock: Lecarrow Clay

A rare pocket of Cenozoic clay, preserved in some large holes or depressions in the local Carboniferous limestone near the town of Lecarrow, gave rise to the once thriving pipe-making industry at Knockcroghery. The clay was (and I say 'was' because much of the good stuff has now gone) a very silica-rich variety, greyish-white in colour, and most suitable for making fired clay pipes.

County Mineral: Melanterite

Roscommon is one of the few counties in Ireland that has supported a working coal mine: that of Arigna. Melanterite is a hydrated iron sulphate mineral that forms in old coal mines (and some metal mines) as a result of iron sulphides oxidising after the mining has ceased. And it so happens that crystals of pale-blue to greenish-blue melanterite, up to 5mm long, have formed on the walls of the old coal mines in the Arigna area.

County Fossil: Gigantoproductus

As the name says, this is a gigantic *Productid* brachiopod that can be found in the Lower Carboniferous limestones of Roscommon, although it isn't exclusive to this county. *Gigantoproductus* is the largest of all the Paleozoic brachiopods, and was probably the largest brachiopod that ever existed. Roscommon specimens regularly reach 15cm in width, which is very big; worldwide, they can reach 30cm in width, which is ... well ... gigantic. Where is the biggest one in Ireland – can you find out?

Sligo

The Coral County

The fossil giant corals on the coast at Streedagh Point and at Serpent Rock (the name is not insignificant) are simply superb; these sites are Irish Geological Heritage sites, so no hammering, please (see Plate Figure 13). What Sligo has here is a wonderfully preserved Carboniferous sea floor, and it is very easy to imagine scuba diving in these tropical shallow seas and stroking, with awe, the forests of giant solitary corals – not to mention all the other fabulous ancient wildlife we can see preserved here. Sligo has something palaeontologically very special.

County Rock: Gabbro

Several counties have gabbro the rock (compositionally similar to basalt but much more coarse-grained, having cooled more slowly within the Earth rather than quickly on the surface), but only Sligo has the Killala Giant Tertiary Gabbro Dyke: the only example in Ireland of a giant gabbro dyke. This special rock occurrence is 400 metres across, and formed due to the partial tearing of Ireland that happened during the Paleocene Epoch (the base of the Paleogene Period) when the North Atlantic was splitting apart. It outcrops on the beach on the west side of Killala, and one can collect lovely minerals that have grown into the open spaces of former gas pockets in the dyke.

County Mineral: Killalaite

As one of the few valid minerals named after an Irish locality, there was no contest for this county mineral. Killalaite is a rare hydrated calcium silicate that was originally described from Killala Bay, near Inishcrone. It is found in some thermally metamorphosed limestones adjacent to two dolerite dykes and occurs as colourless crystals that have a sort of bow-tie shape. Killalaite was likely formed in conditions that were poor in carbon dioxide and at around 475 degrees Celsius.

County Fossil: Siphonophyllia samsonensis

The coral of the Coral County (see Plate Figure 13). A monster coral from the Lower Carboniferous; some specimens are easily 10cm wide and 1 metre in length. *Siphonophyllia samsonensis* are spectacularly displayed at both Serpent Rock and Streedagh Point, the corals possibly having died more or less *in situ* as they were in life – just fallen over a bit.

THE NATIONAL GEOLOGICAL EMBLEMS OF IRELAND

When choosing emblems at the national level, I wanted a combination of geological objects that combine the most characteristic, the most historical, the most unique, and the most visually spectacular geological objects that Ireland has to offer. I have chosen the following as our national geological emblems.

National Rock: Granite

The Irish granites are known the geological world over and have been for generations (see Chapter 12). They are icons and have had an enormous impact on understanding granite generally, constantly being referred to in lectures, books and articles. The Leinster, Donegal and Mourne granites, in particular, hold many 'firsts' between them and have long and distinguished research pedigrees. Furthermore, ongoing research on Irish granites is pointing the way to a more complete understanding of this most enigmatic, yet crucial, rock and its role in the genesis of Earth's continental crust.

National Mineral: Zoned Muscovite

Muscovite is a type of mica that occurs as easily peeled flakes, and is a very common mineral, not just in Ireland but the world. Some of the first observations (possibly *the* first, but to be confirmed) of how it grew were made in 1889 (published 1891) simply by looking at flakes under a polarising microscope, which revealed astoundingly beautiful, quasi-psychedelic growth zones. The growth and corrosion zones are like a mineral version of tree rings and have been used (by yours truly, among others) to make new inferences about the history of the rock from which the muscovite was extracted; muscovite occurs in igneous, metamorphic and sedimentary rocks, so the potential for new information is huge because muscovite is so ubiquitous. And the observations made on Irish muscovites apply to billions of others worldwide. Thus, for historical, aesthetic, scientific and ubiquity reasons, zoned muscovite is Ireland's national mineral. See Plate Figure 9 and this book's front cover for a magnificent zoned muscovite.

National Fossil: Cooksonia

We have already nominated *Cooksonia* as Tipperary's fossil but also select it as our national fossil. It is the world's oldest known vascular plant, from the Wenlock Epoch of the Silurian Period. It was discovered by Offaly native John Feehan and the discovery was published in 1980 in the prestigious science publication *Nature*. It is exceptionally important for the evolution of all plant life on Earth because it is the first plant to grow on land, 425 million years ago. See Plate Figure 12.

14

The Incredible Haughtons – An Irish Earth Science Dynasty

Introduction

Between 1851 and 2015 (and counting), there have been two Haughtons in Ireland who were/are professors of geology and one Haughton who was a professor of geography. All three held their professorships in Dublin and all three came from the same root family of Irish Quakers. They are all related. Between them, they have held academic positions in Ireland for 90 years, with every possibility of reaching the ton in due course. And I have had the pleasure of being taught by two of them.

I am referring to Samuel Haughton, professor of geology at Trinity College Dublin between 1851 and 1881; Joseph Pedlow Haughton, professor of geography at Trinity between 1942 and 1985 (with just one year in Nigeria in the mid-1960s); and Peter David William Haughton, Tullow Oil Professor of Petroleum Geoscience at University College Dublin from 1996 to the present (2015), and who has many years in him yet.

Taking them in chronological order, I offer a portrait of the three Haughtons' significance and contributions to Irish geoscience (and more) and, using genealogical data kindly supplied by Peter Haughton himself, will give the actual family links between all three.

SAMUEL HAUGHTON

All three of our Haughtons are remarkable people. But I think I can say, without offending the other two, that Samuel was particularly remarkable as a polymath of the first order.

Samuel was born on 21 December 1821 in a house beside the River Burrin Bridge, Burrin Street, Carlow town. He was born into a longstanding family of Quakers and Unitarians, although he himself was brought up Church of Ireland. Both his father and his uncle were vehemently anti-slavery.

Samuel was very precocious from the start. While at school, he and his cousin S. Wilfred Haughton built a working steam engine. Samuel entered Trinity at the age of seventeen, although it wasn't geology that he studied but mathematics, in which he obtained the prestigious Gold Medal in his finals. He then took the fearsomely difficult fellowship examinations only a year after his degree and passed first time, despite losing some study time due to a fever. His Church of Ireland upbringing and Trinity position led him to become a priest in 1847; he only rarely actually preached, but he was always a devoutly religious man.

In 1851 he was appointed professor of geology, probably because his interest in optics had led him to an interest in minerals and this was thought good enough by the college board for the position. He remained in the job until he was forced to retire in 1881. In parallel with his work as professor of geology, he studied for and was awarded the degree of Doctor of Medicine in 1862, becoming the chairman of the Medical School in 1879, and was for 34 years on the board of Sir Patrick Dun's Hospital. He was also honorary secretary of the Royal Zoological Society of Ireland, and there is a well-known story of Samuel spending almost two days and nights without sleep dissecting a hippopotamus that had died in Dublin Zoo. Indeed this was but one of some hundred large animals he dissected.

Samuel published papers and books on such diverse topics as mathematics, fluid mechanics, optics, tidal patterns of the Irish seas, animal mechanics, astronomy, and – now almost infamously – how to most efficaciously hang a person so that they wouldn't suffer unduly. Be honest … if you are going to be hanged, that is exactly what you would want!

His geological work was no less prodigious and prestigious. He wrote on all aspects of Irish rocks, minerals, fossils and geological structures. He estimated the age of the Earth, which was a popular topic among Trinity geology professors: two of Haughton's professorial successors, William Sollas and John Joly, also attempted to estimate the age of the Earth. Samuel examined the amount of distortion in fossils to infer the history of the rock as a whole. His anatomical and medical knowledge was used to make deductions about how fossil animals were constructed and how they might have moved. He was among the very first (possibly *the* first) to carry out quantitative chemical analyses of granites or of any igneous rock.

Charles Darwin himself could hardly believe that there was only one Samuel Haughton working out of Dublin, so prolific and diverse was his work. And speaking of Darwin, Samuel, being the religious man that he was, also has the dubious distinction of being the very first person to publicly rubbish, in writing, Darwin's then-new theory of evolution. Nobody's perfect!

Samuel's wife was Louisa Haughton: her father and Samuel's father were half-brothers.

Samuel died on 31 October 1897 as a result of a heart condition. On 3 November, he was given a splendid funeral that went from his home at 12 Northbrook Road East in Dublin, thence into Trinity College for a funeral service in the college chapel, and then a journey by train to his home town of Carlow, where he was finally laid to rest in the family burial ground at Killeshin Church. A fitting end for one of Ireland's greatest polymath virtuosos.

JOSEPH PEDLOW HAUGHTON

Joe, as he is known to colleagues, was born into the Quaker Haughtons. Apparently, he was given a little suitcase by his parents on his very first birthday. And did ever a child get anything so appropriate? Joe became an inveterate traveller at an early age and has never let up. John Andrews, a colleague in the geography department in Trinity, once made the quip that if you were to go to the most isolated spot on Earth and think yourself alone and the first person ever to have set foot there, out of the blue Joe would arrive and probably tell you how it has changed over the last number of years. Joe Haughton is Ireland's geography answer to Sir David Attenborough.

Joe was made professor of geography in Trinity in 1942 and, excepting one year in Nigeria in the mid-1960s, spent his entire academic career in Trinity until his retirement in 1985. His travels to Africa and India were frequent and legendary. As an academic, all his travels had a serious underlying purpose. He was interested in the physical, social, political and economic geography of countries and continents, and all his experiences went straight back into his teaching. This personal element to his geography classes made them come alive in a way that no book-based lecturer could ever hope to emulate. And although a great raconteur with a myriad of stories, he remained astonishingly humble about the knowledge he had gained. He was acutely aware that however much he knew, he was only ever scratching the surface.

Joe taught many generations of geographers over many decades, and many of these geographers subsequently took up academic positions. His legacy and impact is assured.

Joe was invited to join the committees of many local, national and international geographical societies. He was also appointed a council member of the Royal Irish Academy and chairman of the Irish National Committee for Geography, and was the first Irish citizen to be elected to the council of the Institute of British Geographers.

But it is as an enthusiastic world traveller with a sharp observational eye and a scholarly mind who could convey both the subtleties and the big picture of a country, and who could tell you about it from first-hand knowledge, that Joe will have his most lasting impact. Even now, as a nonogenarian, he is *still* travelling and being honoured. To quote from the 2013 news section of Gorta's website:

> Former Professor of Geography in Trinity College Dublin, and ex-Gorta Chair, Joe Haughton, was today honoured in a newly opened Gorta-funded medical and rehabilitation centre in Tamil Nadu, India. The €3.4 million centre will be run by local NGO Social Change and Development (SCAD). Professor Haughton, and his wife Helen, were in India to attend the opening. Haughton, who initiated Gorta's relationship with SCAD in the 1980s, had the administrative block in the new centre named after him. The medical and rehabilitation centre, which was developed in a partnership between Gorta, SCAD and project management company PM Group, will cater for over 1,000 children with mild disabilities.

I realised even as an undergraduate, and more so now, that I was very fortunate to have had geography lessons from Joseph Pedlow Haughton during my first year in Trinity College, 1983–4. The year after teaching me, he retired. But he made an indelible impression. And he was one of those rare men about whom the more you knew and found out the more impressive he became. His knowledge was obvious, yet, as the saying goes, he wore it lightly. And whether it was because he was a Quaker or not, I don't know, but he was incredibly kind and approachable. At the end of one lecture he told the class that he had a few spare copies of the *Atlas of Ireland* (Royal Irish Academy, 1979; now a collector's item) and he would offer them to us for £10 each. Even in 1985, that was ridiculously inexpensive. I wanted one. So off I went the following week with my £10 to his spacious, Victorian-accented, book-walled office where, even though I was a Junior Freshman impinging on his time, he welcomed me in, talked about the atlas, genuinely put me at my ease, and was even thankful that I would consider buying a copy.

Those actions impressed me tremendously: this very distinguished scholar thanking *me* for taking an interest in his work! I left very happy with my new, big green book, in its thick cardboard case, under my arm, and with nothing but admiration for the man. If ever the epithet 'a scholar and a gentleman' be applied to an academic, it should be applied to Professor Joseph Haughton.

PETER DAVID WILLIAM HAUGHTON

The sloughed-off shavings from ancient, weather-scoured mountains. The oozy goo that settles on oceans' bottoms. Splashing about, metaphorically, in oil reservoirs. Scooping sand in the river deltas of the imagination. These are a few of Peter's favourite things. Peter Haughton loves sediments.

Peter was born in Dublin in 1959 and spent his early years there. Between 1970 and 1977 he was sent south to become a boarder at none other than Newtown School in Waterford city – the very same school where I was a pupil between 1969 and 1972: we actually overlapped! Peter entered Trinity College in 1977, graduating in 1981. After doing some mineral exploration work in the Irish Midlands, he went to the University of Glasgow from where he got a doctorate in sedimentology in 1986, figuring out the details of Devonian-age desert sands and their associated transient river deposits and conglomerates, which are now exposed in the Midland Valley of Scotland. He seems to have liked Glasgow academia because he managed to stay on there until 1991. But a 'real job' beckoned and, between 1991 and 1996, he was enticed to work as a consultant in the oil industry. However, he was then enticed back into academia, applying for, and getting, a lectureship as a sedimentologist in the geology department of University College Dublin (now called the UCD School of Earth Sciences). He was now back on home sediment, so to speak, and this is where he remains to this very day, as a professor.

Peter has already made many contributions to understanding Irish geology. He has a talent for going to areas that have been hammered at for many decades and finding something new to say about them. For instance, he has offered new ideas on the original sedimentary environment of the pure quartzites that comprise Bray Head (County Wicklow) – a problem dating back well into the nineteenth century (see Chapter 6 for more). He has also examined in great detail the famous Cliffs of Moher (County Clare) and the whole spectacular sedimentary sequences of rocks that are exposed all along that magnificent coastline. Using inferences from the sedimentary environments in which the original sediments were deposited, he has helped build up a detailed picture of how this part of Ireland looked some 350 million years ago.

My first encounter with Peter was on a Trinity undergraduate field trip to Scotland where one afternoon we were joined by this tall, thin, softly spoken man, who at that stage in his career was finishing up his doctorate at Glasgow University. It was his unenviable task to keep us bunch of students intellectually engaged with the Old Red Sandstone rocks of the northeast part of the Midland Valley of Scotland. We undergraduates could see he knew his stuff, but he spoke so quietly that his voice was often lost in the Scottish wind.

My path crossed his again after he got his lectureship in UCD in 1996 and I, while doing a post-doc in UCD, went to some of his talks. I quickly recognised the same quietly spoken voice, oozing with scholarship, as he showed us meticulous diagrams and offered carefully considered conclusions or possibilities. You always get the feeling with Peter that you are in a safe pair of academic hands. And that is a rare talent, appreciated by students and academics alike.

How Are These Three Professor Haughtons Connected?

Below, and published for the first time, is the relevant portion of the Haughton family tree that shows exactly how all three

Professor Haughtons are related. This is the simplified version, omitting wives, dates of death and so on, to make it easier to see the essential connections. Each generation is numerically labelled, starting at the earliest-traced direct ancestor of them all. Note that, in this scheme, the two '3's of Benjamin and John are the two sons of Isaac. Our three professors are in bold.

1. Wilfred Haughton of Haughton Hall, Cheshire (England)
....2. Isaac Haughton, b. 1663. Moved to Ireland 1694–1700
........3. Benjamin Haughton, of Mullinmast, b. 1705
............4. Samuel Pearson Haughton, of Burrow House (Carlow)
................5. Samuel Haughton, b. 1786
.... 6. **Reverend Samuel Haughton, b. 1821**
........3. John Haughton, of Rheban, b. 1692
............4. Jonathan Haughton, of Ballintore, b. 1726
................5. Joseph Haughton, of Ferns, b. 1766
....................6. Joseph Haughton, b. 1806
........................7. Jonathan Haughton, of Rockspring, b. 1829
............................8. William Haughton, of Coolroe, b. 1850
................................9. Joseph Haughton, b. 1897
....................................10. **Joseph Pedlow Haughton, b. 1920**
....................................10. David Haughton, b. 1921
....11. **Peter David William Haughton, b. 1959**

We can see that the actual common ancestor to all three is Isaac Haughton. Two of Isaac's sons were the brothers Benjamin and John: Samuel is descended from Benjamin, while Joseph and Peter are descended from John. Several websites incorrectly give Reverend Samuel Haughton's father as James Haughton (b. 1795), but it was in fact Samuel Haughton (b. 1786).

Inspection of a standard genealogical canon-law cousin chart reveals all the relationships:

- Peter is Joseph's nephew.
- Joseph is Peter's uncle.
- Reverend Samuel Haughton is Joseph's third cousin four times removed.
- Reverend Samuel Haughton is Peter's third cousin five times removed.

15

Beach Bang Begets Shock Birth

Robert Mallet Makes Earth Move

From your tabloid newspaper reporter, 'Knee-trembler' Johnny O'Shockru, 1859 Special Report

Young Dubliner Robert Mallet has blown up Killiney beach ... and exploded into history. Mallet sent shock waves not only through the sands and rocks of Dublin's premier bathing spot, but through the whole scientific community. He single-handedly invented 'Seismology', the scientific study of earthquakes.

Man of Iron

Robert was born 3 June 1810, in Ryders Row, off Capel Street in Dublin, where his father, John, runs a successful iron foundry business. 'My son has a degree in Natural Philosophy from Trinity, ya know?', says his proud father. 'Robert worked with me for a number of years at the foundry before he went mad blowing stuff up. I thought he was just doing it for the craic, like ... but fair play to him ... he's after making a science out of it', he added.

Rock and Roll

Robert became interested in explosions early in his career, probably when working on various tunnel projects: he was fascinated at the shock waves that the blasting produced and wondered if they could tell him anything about the ground they travelled through.

Shocking

Bob wanted to study these shock waves systematically. If he could do this, he thought, he could nail any relationship between rock type and shock wave. He desperately needed earthquakes by which he could test his 'shocking' new idea. Unfortunately for young Bob – but thank God for the rest of us – Ireland is blessedly free of earthquakes. So 'Bob the Blaster' made his own.

The Big Bang

He jauntily set off to Killiney beach in October of 1849, a beauty spot of unrivalled tranquility in south Dublin, with 11 kegs of gunpowder and an armed guard. There, he dug a huge hole in the shingle into which he put this set amount of gunpowder. Robert then beat a hasty retreat some 500 yards away, where he had set up a device of his own making: a 'seismoscope' as he calls it. This accurately determines when a shock wave has passed by. Robert, though nervous, was now ready to measure all the variables: depth of gunpowder, time of explosion, distance to seismoscope and so on.

Kaboom! The kegs explode. Local auld wans were saying their Hail Marys, thinking it was the end of the world. It was, in fact, the start of a new one. It was the world's first 'controlled source seismic experiment'.

There He Blows!

But he didn't stop there. He blew up other parts of Dalkey. He went over to Wales and blew up bits of that. He was a divil for it. All in the name of science! He was trying to define the properties of the shock waves (earthquakes) in relation to different rocks and structures. And from this he wanted to work back and see from any natural shock waves what the hidden rocks and their structures were. He wanted nothing less than to use these waves as a deep-earth probe. What would he think of next?

Rockin' All Over the World

In 1858, Robert and his son, John (named after Robert's own father), produced an extraordinary map. They rooted through all the literature they could find – and I mean everything, including the Bible – and found every earthquake in history. They plotted all these earthquakes onto a map of the world. To their amazement, and that of their colleagues, they saw that earthquakes occur in particular zones and belts. They are not random. But Robert could not explain the pattern. Your correspondent thinks it might take a hundred years or more before this nut is cracked.

Vocabulary Gets Shaky Start

Have you noticed? New words are coming into everyday speech. Not everyone knows what they mean, but people like the sound of them. Get your tongue around 'seismic', 'seismology' and 'epicentre'. These are words invented by Rob Mallet. Yes – it's a 'seismic shift' in our vocabulary (*groan*, Ed.).

The Man Who Shook the World

Robert Mallet is extraordinary: son of an iron foundry owner; inventor of the science of seismology. And a man unafraid to put the frighteners up the residents of Killiney. Ireland, be proud of 'The man who shook the world'.

16

The Miss Cotter Mystery

Who was Miss Cotter? I don't mean 'What sort of a person was she?', I mean 'What was her name?'

Her full name has been hidden in the historical darkness for over 130 years, even though Miss Cotter achieved everlasting fame as one of the few people to have had a mineral variety named after them. The mineral in question is a variety of arguably the commonest mineral in the Earth's crust: quartz. It is that essential ingredient in sedimentary sandstone, metamorphic quartzite, igneous granite and innumerable mineral veins in which the lucky prospector might find gold. It is also the crunchy stuff that gets into your sandwiches when picnicking on the beach on a windy day. Quartz can be ice-clear or milky-white; in the case of the amethyst variety it can be royal purple, and citrine variety a golden yellow.

But Miss Cotter didn't discover any old quartz. She discovered an extremely rare variety: one that has a unique, pearly metallic sheen, quite unlike any other quartz in the world. And where did she find this? In her back garden – well ... demesne grounds, really – in Rockforest, some three miles east of Mallow, County Cork. In about 1874.

A small limestone quarry had been opened up in Rockforest, and Miss Cotter, evidently an inquisitive type and a lover of nature, went to investigate what was coming out. She discovered a mineral seam that was filled with large, shiny metallic-looking crystals. But they were not of metal. They looked like giant, translucent silver prehistoric hedgehogs that had crawled into a crack and died. She had never seen the like before and contacted the imposingly named Sir Charles Denham Orlando Jephson-Norreys of Mallow Castle. He didn't know what to make of these unusual crystals, either, so he in turn contacted Professor Richard Harkness of Queen's College, now University College Cork. Harkness was a fellow of the Royal Society ... Harkness was a fellow of the Geological Society ... Harkness was going to find out what these crystals really were.

He examined every characteristic he could measure, even whether they withstood being boiled in oxalic acid. He had the public analyst for the City of Cork, Mr Cornelius O'Keeffe, do a full chemical analysis. The prof concluded, in a paper published in 1878, that the strange silvery mineral was a unique variety of quartz, to be hereafter called cotterite after its discoverer, Miss Cotter. In print, he only ever called her 'Miss Cotter'. The silvery pearly sheen was apparently due to a last-gasp layer of quartz-plus-ultrafine-clays that had grown on top of otherwise normal quartz. By 1877, no more cotterite was to be found.

However, Miss Cotter and her elderly father, the Reverend George Edmund Cotter, both acted to make this unusual style of quartz as well known as possible. Specimens were sent to the Royal Dublin Society, and to the British Museum of Natural History. There are three specimens in the National Museum of Ireland – Natural History, and two in the collections of Trinity College Dublin.

Genuine cotterite remains one of the very rarest varieties of quartz. In Ireland, we have the biggest specimen in the world. This is on public display in the Cork Geological Museum in University College Cork. Go and see it! It is awe-inspiring. And a true national treasure.

But ... who was Miss Cotter? Pulling together every clue at my disposal – and making excellent use of a birthday gift subscription to the Irish genealogical website Findmypast.ie – I managed to find out. And I can now reveal her name: a name not found in any publication, Google search, or Wikipedia entry on cotterite.

Grace Elizabeth Cotter. Born in 1830 to Reverend George Edmund Cotter and Grace La Touche. She died, unmarried, in the same year as her father (1879), aged 49: only one year after her name was *not quite* immortalised by Professor Harkness.

Miss Grace Elizabeth Cotter: the discoverer of cotterite. A rare Cork woman who gave her name to a rare bit o' stone.

17

GEOLOGY – THE FATAL ATTRACTION

Geology can be a thrilling career, living life on the edge and in the wilds. If one takes due care, it is rare for something to go badly wrong. But once in a while things do go wrong, and the life of the practising geologist can be cut short. The Irish geological community is like that of a small village – we all know, or know of, each other. So, when a member of that tight-knit bunch dies while doing geological work, it is keenly felt. Geology is a human activity, after all. The three geologists below all died young while actively engaged in the subject they loved. Two of them I knew personally.

DAVE JOHNSTON

I was in my second year at Trinity College Dublin (1984–5) when, on the balcony of the geology department, in the Museum Building, I got talking to a young, freshly minted PhD who had just been appointed the department's new structural and economic geology lecturer. He was stocky, quietly confident, had a bit of a swagger and wore a big smile on his stubbly, square-cut face. He had a unique accent: a cross between

Northsider Dublin and Australian. Dave Johnston was a former graduate of Trinity, obtaining a first-class honours degree in 1980, and he had done his doctoral research in Monash University in Melbourne, where he picked up not only an Australian accent but also innumerable jokes and songs (which on field trips he made absolutely sure we all knew!). I remember him saying to me that he had not been Trinity's first choice for the job, but when that first-choice candidate bailed out, the selection board chose Dave. And an excellent choice they ultimately made.

From the first, Dave dived headlong into the Irish geological scene and continued as he had started throughout his all-too-short career: giving lectures and talks on his Australian adventures to professionals and amateur groups alike; initiating research projects, such as applying a new branch of mathematics (fractal geometry) both to structural geology problems and to industrial metal deposits, this latter application allowing one to make more accurate mineral reserve estimates; discovering new mineral textures, such as ikaite pseudomorphs in Donegal and deducing that this was additional evidence for Ireland being in the grip of an ice age some 500 million years ago; and finding out that the 3D shape of what geologists call 'en echelon' veins is like that of a ship's propeller. He was an active member of the Irish Geological Association, acting as president in 1993 and 1994, and was a council member of the Irish Association for Economic Geology.

Dave most certainly did have a quiet side, and he could show disarming personal sensitivity. He could also be very boisterous. And though he was relatively relaxed about student behaviour on field trips, one thing he would not tolerate was littering the countryside: any student caught doing that would get a good bollocking. His breadth of knowledge and his principle of having fun in a way that harmed neither the landscape nor other people meant that Dave was respected and sought out by students, landowners and fellow lecturers alike.

He loved to socialise and people loved to be around him. Summer barbeques were often held in the back garden of his Killester (Northside Dublin) home and, with potentially dozens of senior undergraduate and postgraduate students invited, these could become rowdy affairs. For example, the garden hedge was accidentally set alight one year and the Dublin Fire Brigade had to be called. Bless his neighbours: they were very understanding.

It almost goes without saying that Dave was very smart and was held in much admiration by fellow academics. Stephen Daly (UCD's School of Earth Sciences) once said that Dave, or 'Johnners' as he called him, was the only person he knew who could go out for a few pints of an evening then go back to the lab and still do useful geological research.

The last year of his life was, to the dismay of those of us looking on from the sidelines, not a happy one. He lost his beloved father and also endured the emotional turmoil of having his marriage fall apart. The ultimate tragedy, however, came on a blustery Monday, 2 October 1995, at Annagh Head in northwest County Mayo. Nobody witnessed the actual event, but it seems that Dave was studying, probably too intently, the complex gneiss rocks on a wave-cut platform along that stretch of coast when a freak wave hit the shores and dragged him out to sea. And there he presumably drowned.

I say 'presumably' because his body was never recovered. The sea still holds that secret. I remember the visceral shock that went through the Irish geological community when the news filtered through. Within a year, there was a special issue of the *Irish Journal of Earth Sciences* in his honour (Vol. 15, 1996), and over the weekend of 8–9 November 1997, Trinity College hosted a full international colloquium, the Dave Johnston Memorial Meeting. Dave was a dynamic, intelligent and generous character and, to those who knew him, he remains painfully missed.

Thorley Mark Sweetman

Anyone doing geological work in the Blackstairs Mountains of south Leinster must read Thorley's work. His papers and doctoral thesis on the granite and the metamorphic rocks of the region are high-quality productions. I didn't know Thorley personally, although I saw him at various geology meetings in the mid-to late 1980s. I do know that he was universally loved and held in great regard.

Thorley Mark Sweetman was born in the first half of 1957 in Sheffield, Yorkshire, to Ernest V. Sweetman and Agnes Threlfall.

He was evidently very active and talented as a teenager because, in 1974, he became one of only a very few to be awarded the highest honour that can be awarded a boy scout: the Queen's Scout Award, which he gained from his time with the Woodseats Venture Unit in south Sheffield. His name is writ large in gold letters on the Woodseats scout troops' honours board at its headquarters.

After he completed his undergraduate geology degree at the University of Sheffield in 1979, he arrived in Ireland to start a doctorate under the supervision of Professor Peter Brück at University College Cork (UCC) on the geology of the Blackstairs Mountains. The area had been mapped only once before by staff of the Geological Survey of Ireland in the 1850s. Thorley became only the second person to map it in some 130 years! But it was worth the wait. In January 1984, he produced a magnificent two-volume PhD thesis titled 'The Geology and Geochemistry of the Blackstairs Unit of the Leinster Granite', which included stunning hand-coloured detailed geological maps, exceptional old-style pen-and-ink drawings of rock textures, photographs of all the key features that he observed, and the very first systematic set of chemical and isotopic analyses of the granites from this area. Thorley published several important papers from this thesis and gave us a modern viewpoint from which to discuss these granites and their associated rocks.

He did such a good job, in fact, that he was appointed a geology lecturer in UCC and remained in that post from 1985 to 1989. By this stage, he had married Marian, an art teacher from Mayfield Community School in Cork, and they would often go on various adventures together when time allowed. On one occasion in 1986 they went mountaineering in the Himalayas and happened to come across one Derek Jackson, who was the chief leader of the British Schools' Exploring Society. Derek was suffering badly from diarrhoea and Marian managed to cure him. As a result, Thorley was invited to be chief scientist on the society's 1988 summer expedition, with Marian also invited, plus two local Irish lads. These are the benefits of being a Good Samaritan, as the *Cork Examiner* newspaper was to put it.

However, Thorley was not to stay in Ireland. In April 1989, he and Marian moved to the University of Zimbabwe, in Harare, where Thorley worked as a lecturer and researcher on the geochemistry of the ancient Precambrian rocks there. And it was while collecting samples in the field, working in the Mana Pools area of Zimbabwe, on 24 August 1991, that Thorley's life came to a sudden and unexpected end. He was killed, trampled to death, by an elephant. He was only 34. His grave is in Zimbabwe. Marian moved back to Ireland.

Thorley left an indelible, positive impression on the geology department of UCC during his time there, as both a postgraduate and a lecturer. And in his honour, UCC – through contributions from his colleagues, students and family – set up the Thorley Sweetman Prize for the best BSc (Honours) undergraduate mapping project. The winner gets a substantial cash award, a parchment, and his or her name inscribed on a trophy. But anyone who has seen Thorley's own maps will realise that there is an incredible standard to live up to.

Postscript: it so happened that I was interviewed in London for Thorley's Zimbabwe post after he was killed when the lectureship position became vacant. Despite being offered the job, I declined.

George Mooney

George Mooney and I hit it off immediately from the moment we sat next to each other in our first geology lectures in Trinity College Dublin in October 1983. We had a lot in common: an interest in geology since childhood; doing the Trinity geology matriculation exam for those vital extra Leaving Certificate points; and going to college expressly to study geology. Both of us smoked roll-your-own cigarettes. But there were enough differences to make life interesting. George was more 'on edge' than I was and his fingernails were permanently chewed to the quick; he had an encyclopaedic knowledge of Monty Python. Sometimes, when I see

sheep in a field I chuckle when I recall his brilliant imitation of the immortal line: 'They do not so much fly, as plummet'. George was a great fan of the band The B-52's, his fingers often 'playing' the riff to their song 'Rock Lobster' in the air. He also had a subscription to *Scientific American*.

George and I were close friends throughout our four undergraduate years and we did a lot of student mapping projects and practical laboratory work together. Over time, we developed an element of 'old married couple' syndrome: we didn't always agree – George was quicker with opinions and firmer in his views than I was. But we never fell out.

By far the most physically and psychologically demanding thing we both did was to spend the best part of two months during the summer (the *what*?) of 1986 on Gorumna Island, and the nearby Lettermore and Lettermullen islands, for our senior geology-mapping projects. Back then, every undergraduate geology student had to spend at least six weeks in the wilds of Ireland during the summer between their third and fourth years at university to produce a geological map of an area of their choice. This map would be assessed by Trinity's geology staff and count towards one's fourth-year final degree mark.

Gorumna and the other islands are at the edge of Galway Bay, practically in the Atlantic, and are blessed with excellent geological variety. But their location means they are relatively barren places and are totally exposed to any storms rolling in. We were not to know it, but this was the year of Hurricane Charlie. Our mode of transport was the bicycle; our accommodation was in tents.

George had camped before, but I had not. He had the foresight to bring an inflatable comfy mattress to put inside his tent. I, on the other hand, had a thin sleeping bag that I placed directly on the ground. Worse, I had contrived to set up my tent on a slope, on some uncomfortable rocks and on top of a ridiculously robust thistle. A comprehensive account of what we suffered would fill a short book. Neither of us saw the sun *at all* for the first two weeks we were there. Our food budget was very tight and, as a result, every day we ate the same thing: cornflakes for breakfast, one marmalade sandwich for lunch. Evenings were spent fighting the wind to get the portable gas stove to light and so heat a half-can portion of beans and half-can portion of corned beef each. When we could afford it, we cycled to the local pub in the evening for a bit of warmth and to (very slowly) consume one pint of Guinness, one packet of cheese and onion crisps and one Mars bar.

But, for weeks on end, the wind and the rain were utterly relentless. The wind drove the rain through our tent fly sheets so hard that inside the tent was like sitting in an endlessly noisy, flapping, cold shower. Everything was permanently cold and wet. This was supposed to be summer! On one occasion, I went 72 hours without sleep; another time I woke up with my feet, still in my sleeping bag, in three inches of water that had ponded at the bottom of my sloping tent.

During one of the Force 9 gales, George and I struggled down to the pub of an evening to see the TV weather forecast for the next day … and just laughed. We were facing impending disaster. The next 24 hours promised a full-on Force 10 storm, heading directly towards us. I filled my tent with large boulders borrowed from the local stone wall … but George didn't. And the inevitable happened. At about 2.00 a.m. in the pitch dark, at the worst possible time, the screaming winds ripped George's tent apart, bent and twisted the framework metal poles like spaghetti and sent the tent pegs whizzing off to kingdom come. Following George's shouts for help, we both ran around in the dark trying to salvage his belongings, which were being scattered everywhere. And in the maelstrom we could just make out the bat-like remains of his tent being tossed into the local lake, some 100 feet away.

After that, we both went back to my tent, crouching between the boulders, and just talked the rest of the night away, George dryly noting that I had no comfy bed. Red-eyed at about 7.00 a.m. from no sleep and with the wind still roaring we decided to skip that day's mapping. Instead, we both cycled off to an old ruin at the end of Lettermullen Island to have a smoke and cathartically watch the huge waves coming in.

After being awarded his degree from Trinity, George got a job with a South American mining company that was prospecting for gold in the region's mountainous jungles. George was part of the exploration team and was regularly involved in doing large-scale reconnaissance work in a light aeroplane. One day, not long after he had joined the company – and not long after he had developed a wonderful relationship with his new South African girlfriend, Annerley, and her child from a previous relationship (thereby making George a dad overnight) – the light plane he was in crashed into the side of a mountain, killing George instantly.

18

EMERALD'S ISLE

This short story rather beautifully combines the themes of gems, genealogy and Ireland.

When researching family history, one has to occasionally grapple with surname variants. My own include Roycroft, Roycraft, Raycroft, Recroft, and many others. One such variant not (yet) connected to my own Roycroft family (who have nested in West Cork for several centuries as tenant farmers, possibly as far back as 1587) is the surname 'Rowcroft'. This variant is very rare in Ireland. But there is one Rowcroft family that may be unique and in a most unexpected way.

Ireland contains no indigenous emeralds. Yet Ireland does hold one emerald to its warm green heart – Emerald Rowcroft. She was born between January and March 1890 in Ballycastle, County Antrim, to George Vernour Rowcroft (1857–1908) and Kathleen (or Catherine) Olivia Boyd (1859–1945). Kathleen's own parents were General Hugh Boyd and Frances Millicent Dobbs, and through her Dobbs line Kathleen had deep Antrim roots. George Vernour Rowcroft was born in India to Francis Frederick Rowcroft and Sarah Johnstone Souper and was subsequently sent to England, where he was brought up first by his grandparents and later, as a teenager, by his uncles and aunts. It's a characteristic of this entire family that they never stayed in one place for any length of time.

After marrying Kathleen on 12 November 1885, in London, she and George moved to Birtle in Marquette, Manitoba, Canada.

During their stay in Antrim and Manitoba, respectively, they produced four children: two were born in Manitoba and two in Antrim. And just look at the names of these four children:

Pearl Rowcroft (daughter, born 1887 in Canada)
Beryl Rowcroft (son, born 1888 in Canada)
Emerald Rowcroft (daughter, born 1890 in Ballycastle, Antrim)
Garnet Rowcroft (son, born 1895 in Ballycastle, Antrim)

George and Kathleen named all four of their children after gemstones. Although they went by those names (and even appear on census forms with those names), their full names, to be totally precise, were Agnes Millicent Pearl Rowcroft, Beryl Vernour Noel Rowcroft, Kathleen Emerald Rowcroft, and Hugh Garnet Crichton Rowcroft. And yes, 'Beryl' here is a male name – just check his army records.

This familial geological theme appeared elsewhere in the Rowcroft family: George's brother (a major in the Indian army) was Ernest Cave Rowcroft; and the daughter of George Francis Rowcroft and Florence Marion Eva Hennessy (George Francis being the son of George Cleland Rowcroft, the brother of Francis Frederick Rowcroft) was Ruby Frances Rowcroft, making her a second cousin to our four gems above.

Tragically, Baby Emerald died in 1891 in Ballycastle when she was just one year old. Four years later, Garnet was born, after which George and the family returned to Canada where tragedy struck again in 1908. This time it was George himself who died, aged a relatively young 45. This left Kathleen alone to bring up the three remaining solitaires, so they all moved yet again, this time to England, where Pearl and Beryl could now take care of themselves. Garnet, however, was sent to boarding school in Bristol (he was there in 1911), and subsequently managed to survive the horror of fighting in World War I.

During the years following the family's arrival in England, life took a turn for the better. Pearl, Beryl and Garnet thrived as adults and had, from what I can deduce, good lives, as indeed did cousin Ruby. Having evidently inherited the family's itchy feet, Garnet upped sticks and emigrated to Australia in 1925, as a farmer, moving house a number of times while there. He passed away in

Brisbane in 1968. Pearl, despite being the firstborn, outlived everyone, passing away at the grand age of 89 in 1976.

But if you are ever in Ballycastle, spare a thought for Emerald. She has no close family near her. Her mother and father, and even her grandparents, led lives that took them to very distant places. Her fellow gem-named brothers, sisters and cousins, whom she never had a chance to get to know, grew up and scattered to the four corners of the earth. But here in Ballycastle, the Emerald Isle will forever be Emerald's isle.

19

648 Billion Sunrises: In the Beginning

Hiberno-Genesis: The Book of Rocks

And lo, it was written ...

1. In the Precambrian, Geology created the continents and the oceans.
2. Now, Ireland was formless and empty and was as two separate halves. And the halves did reside in the larger northern land of Laurentia and in the smaller southern land of Avalonia. And the waters between the lands were the Iapetus Ocean.
3. And Geology said, 'Let the continents collide,' and the larger and the smaller lands did obliquely collide. And the earth did buckle and melt.
4. And Geology made the collision, and it was good. And Ireland was separate no more.
5. Geology called the new land 'Ireland' that was formed from Laurentia and accreted Avalonia. And Ireland, not yet an island, saw its first evening and morning.
6. Geology said, 'Let there be a Suture between the old lands to weld what was once Laurentia and Avalonia.'
7. So, Geology made the Iapetus Suture. And the weld was good.
8. And Geology said, 'Let the seas of the Iapetus dry and let mountains arise.'
9. And Geology called these mountains 'The Caledonides', and they stretched for a thousand kilometres northeast and southwest.

10. And Geology said, 'Let the land bring forth the first plant at Kiltorcan and the four-footed beast at Valentia.'
11. And the land produced the plants and the beasts according to their kinds, and lakes did form and rivers did run. And in the evenings and mornings, granite did uplift and erode.
12. And Geology said, 'Let there be little atoms that do decay and let them serve to mark the passing of the ages in thousands and yet also in millions of years.' And it was so.
13. And the atoms did knowingly decay. And the atoms were measured, and the ages were known. And it was good.
14. And Geology did make a second great continent to collide. Gondwana did collide with the sutured Caledonides and so it was that the Variscide Mountains were formed. And there was born the great supercontinent of Pangaea.
15. And there was a rumpling in the south of Ireland.
16. And yet, Ireland was still not an island, but was enclosed within the folds of the great supercontinent Pangaea.
17. So Geology created the deserts and the dinosaurs and the warm waters that teemed with living things, and the first mammals and the flowers and the winged reptiles.
18. But Geology did strip the land of Ireland of its dinosaurs and of its rocks so filled with teeming living things.
19. And much mystery did surround the Mesozoic Era of Ireland.
20. The mighty Pangaea itself then did split, and waters filled the land. And Geology did call the new waters 'the Atlantic Ocean', and America and Eurasia did rift and drift.
21. And basalts flooded the land, and the land did tear asunder in terrible wounds; rock ripped rock and magma ran red.
22. But lo! Ice and snow did come and great beasts did live in the freezing torment. And Man came and did live off the beasts.
23. Man and beast suffered but a short time. Geology warmed the land. And verdant forests did cover Ireland, and the creatures and the livestock there did flourish.
24. And Geology only now did Ireland an island make, according to its shape. And it was good.
25. And Geology saw the island it had made and saw that it was good. And in the evening and morning Man rejoiced.
26. And it was very good.

REFERENCES

This book is, at heart, a miscellany of Irish geology. It is based on information taken from authoritative texts, articles and websites, as well as personal knowledge and some original and/or unpublished research, notably for Chapters 2, 14, 16, 18 and a lesser amount for Chapters 10, 11 and 12. It was a deliberate policy not to include too many references in the text. To indicate my main sources, however, and to save space, I will give a general list below of works that I consulted extensively for many of the essays, and then give a list (not exhaustive) of more specific individual sources not covered in the general list. The interested reader can readily search the internet (using authoritative sites) to look up any additional word, concept, or picture of a dinosaur/fossil/mineral.

GENERAL REFERENCES

Holland, C.H. and Sanders, I.S. (eds) (2009), *The Geology of Ireland* (second edition), Edinburgh: Dunedin Academic Press.

Kennan, P. (1995), *Written in Stone,* Dublin: Geological Survey of Ireland.

Lyle, P. (2003), *The North of Ireland,* Classic Geology in Europe 5, Harpenden: Terra Publishing.

Meere, P., MacCarthy, I., Reavy, J., Allen, A. and Higgs, K. (2013), *Geology of Ireland: A Field Guide,* Cork: The Collins Press.

Mulvihill, M. (2002), *Ingenious Ireland: A County-by-County Exploration of Irish Mysteries and Marvels,* Dublin: Townhouse Press.

Nayor, D. and Shannon, P.M. (2011), *Petroleum Geology of Ireland,* Edinburgh: Dunedin Academic Press.

Sleeman, A.G., McConnell, B. and Gatley, S. (2004), *Understanding Earth Processes, Rocks and the Geological History of Ireland: A Companion to the 1:1,000,000 Scale Bedrock Geological Map of Ireland*, Dublin: Geological Survey of Ireland.

Stillman, C. and Sevastopulo, G. (2005), *Leinster*, Classic Geology in Europe 6, Harpenden: Terra Publishing.

Tindle, A. (2008), *Minerals of Britain and Ireland*, Harpenden: Terra Publishing.

Wyse Jackson, P., Parkes, M. and Simms, M. (2010), *Geology of Ireland: County by County*, Dublin: Geoschol (Trinity College Dublin).

Woodcock, N. and Strachan, R. (eds) (2002), *Geological History of Britain and Ireland*, Oxford: Blackwell Publishing.

I have made extensive use of the Geological Survey of Ireland's complete series of booklets (with accompanying maps) that describe the geology of all the 1:100,000 bedrock maps of Ireland, specifically the following: Sheet 1 (and part of sheet 2), 3 (and part of sheet 4) [There is no sheet 5], 6, 7, 8 (and part of sheet 9), 10 (both versions), 11, 12, 13, 14, 15, 16, 17, 18, 19, 20, 21, 22, 23, 24, 25.

I also made extensive use of the complete set to date (1978–2014) of the *Irish Journal of Earth Sciences*.

Stephen Moreton, a professional chemist based in northwest England and an exceptional amateur mineralogist who collects and publishes on Irish minerals, very generously supplied me with information concerning semi-precious mineral occurrences in Ireland.

Selected References

Abstracts of 57th Irish Geological Research Meeting, 28 February–2 March 2014, University College Dublin (and my own notes taken during talks at that meeting).

Anonymous, no year, *Carlow County Geological Site Report*.

Baxter, S. (2011), *A Geological Field Guide to Cooley, Gullion, Mourne and Slieve Croob*, Louth County Council.

Book of Genesis, *The Bible* (inspiration for Chapter 19).

Brück, P.M. and O'Connor, P.J. (1977), 'The Leinster Batholith: Geology and Geochemistry of the Northern Units', *Geological Survey of Ireland Bulletin*, Vol. 2, pp. 107–41.

Brunton, H. (1966), 'The Morphology of *Delepinea destinezi* (Vaughan) (Brachiopoda: Daviesiellidae)', *Annals and Magazine of Natural History*, series 13, Vol. 9, Nos. 103–5, pp. 439–44.

Capper, B.P. (1829), *Topographical Dictionary of the United Kingdom*, London: Sir Richard Phillips and Co.

Clarkson, E.N.K. and Tripp, R.P. (1982), 'The Ordovician Trilobite *Calyptaulax brongniartii*' (Portlock), *Transactions of the Royal Society of Edinburgh: Earth Sciences*, Vol. 72, pp. 287–94.

Clemens, J.D. and Mawer, C.K. (1992), 'Granitic Magma Transport by Fracture Propagation', *Tectonophysics*, Vol. 204, pp. 339–60. (And see the commentary on the reprint of this article in J.D. Clemens and F. Bea (eds), Chapter 10, *Landmark Papers Number 4, Granite Petrogenesis*, 2012, pages L117–L118, Mineralogical Society of Great Britain and Ireland.)

Critchley, M. (2002), 'A Survey of Tankardstown Mine, Bunmahon, Co. Waterford', *Journal of the Mining Heritage Trust of Ireland*, Vol. 2, pp. 21–4.

Daly, S.J., Muir, R.J. and Cliff, R.A. (1991), 'A Precise U–Pb Zircon Age for the Inishtrahull Syenitic Gneiss, Co. Donegal, Ireland', *Journal of the Geological Society, London*, Vol. 148, pp. 639–42.

Dana, E.S. (1892), *The System of Mineralogy of James Dwight Dana (1837–1868)*, sixth edition, London: Kegan Paul, Trench, Trübner and Co.

Feehan, J. (2013), *The Geology of Laois and Offaly*, Offaly County Council.

Fitton, W. (1812 [based on an earlier 1811 paper by Fitton]), *Notes on the Mineralogy of Part of the Vicinity of Dublin: Taken Principally from Papers of the late Rev. Walter Stephens, A.M.*, London: William Phillips and George Yard.

Giesecke, C.J. (1832), *A Descriptive Catalogue of a New Collection of Minerals in the Museum of the Royal Dublin Society to which Is Added an Irish mineralogy*, Dublin: Graisberry.

Greg, R.P and Lettsom, W.G. (1858), *Manual of the Mineralogy of Great Britain and Ireland*, London: John van Voorst.

Hall, A. (1972), 'New Data on the Composition of Caledonian Granites', *Mineralogical Magazine*, Vol. 38, pp. 847–62.

Harkness, R. (1878), 'On Cotterite, a New Form of Quartz', *Mineralogical Magazine*, Vol. 2, pp. 82–4.

Hatch, W.H. (1888), 'On the Spheroid-Bearing Granite of Mullaghderg', *Quarterly Journal of the Geological Society of London*, Vol. 44, pp. 548–60.

Haughton, S. (1856a), 'Experimental Researches on the Granites of Ireland: Part I – On the Granites of the South-East of Ireland', *Quarterly Journal of the Geological Society of London*, Vol. 12, pp. 171–202.

Haughton, S. (1856b), 'On the Chemical Composition and Optical Properties of the Mica of the Dublin, Wicklow and Carlow Granites', *Proceedings of the Royal Irish Academy*, Vol. 6, pp. 176–9.

Haughton, S. (1856c), 'On the Granites of the Province of Leinster', *Proceedings of the Royal Irish Academy*, Vol. 6, pp. 230–2.

Haughton, S. (1858) [the second part of a joint paper by J. Beete Jukes and Haughton], 'On the Lower Palaeozoic Rocks of the South-East of Ireland, and their Associated Igneous Rocks', *Transactions of the Royal Irish Academy*, Vol. 23, pp. 563–622.

Joly, J. (1917), 'The Genesis of Pleochroic Haloes', *Philosophical Transactions of the Royal Society of London*, Vol. 217, pp. 51–79 (earlier references by Joly on the subject are here).

Jones, M. (1994), 'Coticule and Related Rocks: Their Significance in the Caledonian-Appalachian Orogen', PhD thesis, National University of Ireland.

Kane, R.J. (1844), *The Industrial Resources of Ireland*, Dublin: Hodges and Smith.

Kennan, P.S. (1997), 'Granite: A Singular Rock', John Jackson Lecture 1997, *Occasional Papers in Irish Science and Technology*, No. 15.

Lamont, A. (1941), 'Trinucleidae in Eire', *Annals and Magazine of Natural History*, Series 11, Vol. 8, No. 47, pp. 438–69.

Lamplugh, G.W., Kilroe, J.R., M'Henry, A., Seymour, H.J. and Wright, W.B. (1903), 'The Geology of the County around Dublin', *Memoirs of the Geological Survey of Ireland*, Dublin: Alex Thom and Co.

Lewis, S. (1837), *Topographical Dictionary of Ireland*, London: S. Lewis.

Lucey, J. (2005), *The Irish Pearl: A Cultural, Social and Economic History*, Bray: Wordwell Limited.

McArdle, P. (2011), *Gold Frenzy*, Swinford: Albertine Kennedy Publishing.

McCaffrey, K. (1996), 'Dave Johnston: An Appreciation', *Irish Journal of Earth Sciences*, Vol. 15, pp. 1–3.

Mohr, P. (1991), 'Cryptic Sr and Nd Isotopic Variation across the Leinster Granite, Southeast Ireland', *Geological Magazine,* Vol. 128, No. 3, pp. 251–6.

Molony, L. (2013), 'Amber in Prehistoric Ireland', *Archaeology Ireland,* Vol. 27, No. 1, pp. 13–16.

Oliver, G.J.H. (1978), 'Prehnite–pumpellyite Facies Metamorphism in County Cavan, Ireland', *Nature,* Vol. 274, pp. 242–3.

Orr, P.J. and Briggs, D.E.G. (1999), 'Exceptionally Preserved Conchostracans and Other Crustaceans from the Upper Carboniferous of Ireland', The Palaeontological Association, Special Papers in Palaeontology, Vol. 62, pp. 1–68.

Orr, P.J., Briggs, D.E.G. and Kearns, S.L. (2008), 'Taphonomy of Exceptionally Preserved Crustaceans from the Upper Carboniferous of Southeastern Ireland', *Palaios,* Vol. 23, pp. 298–312.

Owen, R. (1861), 'A Monograph of a Fossil Dinosaur (*Scelidosaurus harrisonii* Owen) of the Lower Lias', *Palaeontological Society Monographs,* Part 1, pp. 1–14.

Parsons, I. (2013), 'A Record Brimful of Promise', *Elements,* Vol. 9, No. 1, p. 79.

Pitcher, W.S. (1997), *The Nature and Origin of Granite,* second edition, London: Chapman and Hall.

Pitcher, W.S. and Hutton, D.H.W. (2003), *A Master Class Guide to the Granites of Donegal,* Dublin: Geological Survey of Ireland (with map).

Reynolds, D.L. (1943), 'Granitisation of Hornfelsed Sediments in the Newry Granodiorite of Goraghwood Quarry, Co. Armagh', *Proceedings of the Royal Irish Academy,* Vol. 48 (section B), pp. 231–67.

Richey, J.E. (1928), 'Structural Relations of the Mourne Granites (Northern Ireland)', *Quarterly Journal of the Geological Society of London,* Vol. 83, pp. 653–88.

Richey, J.E. (1932), 'The Tertiary Ring Complex of Slieve Gullion (Ireland)', *Quarterly Journal of the Geological Society of London,* Vol. 88, pp. 776–849.

Romano, M. (1980), 'The Stratigraphy of the Ordovician Rocks between Slane (County Meath) and Collon (County Louth), Eastern Ireland', *Irish Journal of Earth Sciences, Royal Dublin Society,* Vol. 3, pp. 53–79.

Roycroft, P.D. (1995), 'Zoned Muscovite: Description and Applications to the Leinster Granite and Other Muscovite-Bearing Rocks', PhD thesis, National University of Ireland.

Ryback, G and Francis, J.G. (1991), 'Microminerals from Ireland, Part 1: The Southwest (Munster)', *UK Journal of Mines and Minerals*, Vol. 10, pp. 22–7.

Simms, M.J. and Monaghan, N. (2008), 'The Origin and Occupation History of Poll na mBéar, Glenade, Co. Leitrim', *Irish Speleology*, Vol. 17, pp. 50–3.

Sollas, W.J. (1891 [written in 1889]), 'Contributions to a Knowledge of the Granites of Leinster', *Transactions of the Royal Irish Academy*, Vol. 29, pp. 427–514.

Standish, C., Dhuime, B., Chapman, R., Coath, C., Hawkesworth, C. and Pike, A. (2013), 'Solution and Laser Ablation MC–ICP–MS Lead Isotope Analysis of Gold', *Journal of Analytical Atomic Spectrometry*, Vol. 28, pp. 217–25.

Standish, C.D., Dhuime, B., Hawkesworth, C.J. and Pike, A.W.G. (2015), 'A Non-Local Source of Irish Chalcolithic and Early Bronze Age Gold', *Proceedings of the Prehistoric Society*, doi:10.1017/ppr.2015.4. (I have also used notes taken at a lecture that Chris gave in UCD in 2014 in which he presented his new data and ideas.)

Sweetman, T. (1988), 'The Geology of the Blackstairs Unit of the Leinster Granite', *Irish Journal of Earth Sciences*, Vol. 9, No. 1, pp. 39–60.

Taylor, T. (1818), 'An Account of a New Mineral Substance, Discovered at Killiney, in the Vicinity of Dublin', *Transactions of the Royal Irish Academy*, Vol. 13, pp. 3–11.

Tilley, C.E. and Vincent, H.C.G. (1948), 'The Occurrence of an Orthorhombic High-Temperature form of Ca_2SiO_4 (bredigite) in the Scawt Hill contact zone and as a Constituent of Slags', *Mineralogical Magazine*, Vol. 28, pp. 255–71.

Visschler, H. (1971), 'The Permian and Triassic of the Kingscourt Outlier, Ireland: A Palynological Investigation Related to Regional Stratigraphical Problems in the Permian and Triassic of Western Europe', Geological Survey of Ireland Special Paper No. 1.

Walsh, S. (2014), *Connemara Marble: Ireland's National Gem*, Dublin: O'Brien Press.

Walshe, J. (2013), Unpublished Fieldtrip Notes to Loughshinny, County Dublin, University College Dublin.

Weaver, T. (1819), 'Memoir on the Geological Relations of the East of Ireland', *Transactions of the Geological Society of London*, Vol. 5, No. 1, pp. 132–269.

White, M. (1992), *Mount Leinster: Environment, Mining and Politics*, Dublin: Geography Publications.

Zhang, S., Li, Z.-X., Evans, D.A.D., Wu, H., Li, H. and Dong, J. (2012), 'Pre-Rhodinia Supercontinent Nuna Shaping Up: A Global Synthesis with New Paleomagnetic Results from North China', *Earth and Planetary Sciences*, Vols. 353–4, pp. 145–55.

Genealogical Note: I have used Findmypast.ie (world subscription) and many free genealogical websites (e.g. FamilySearch.org) in the Miss Cotter, Thorley Sweetman and Emerald Rowcroft essays. Untangling the Rowcrofts, for example, was a lesson in not believing most online family trees and double-checking everything. If you, the reader, want to verify the Rowcroft genealogy, be aware that very often the father of George Vernour Rowcroft is given as George Cleland Rowcroft (instead of the correct Francis Frederick) and there is confusion with Beryl Rowcroft and Ruby Rowcroft, mistaking both for females when Beryl is actually male. Sometimes one even finds names like Beryl Ruby Frances Rowcroft, which disturbingly mixes up both cousins and genders!

Index

Illustrations in the main text are indicated by page numbers in italics. Plate figures are denoted by 'pf' followed by the figure number.